Generis

PUBLISHING

INFORMATION COMMUNICATION TECHNOLOGY PREPAREDNESS

INTEGRATION IN EDUCATION AND STAKEHOLDERS' PERCEPTIONS ON CHEMISTRY PERFORMANCE AMONG PUBLIC SECONDARY SCHOOLS IN KISUMU COUNTY, KENYA

Ruth Atieno Otiang'a

CIP a Camerei Naţionale a Cărţii

Otiang'a, Ruth Atieno.

Information communication technology preparedness : Integration in Education and Stakeholders' perceptions on Chemistry performance among Public Secondary Schools in Kisumu County, Kenya / Ruth Atieno Otiang'a. – Chişinău : Generis Publishing, 2020 (Print on demand). – 188 p. : tab.

Referinţe bibliogr.: p. 134-155.

ISBN 978-9975-153-81-2.

37.091:004(676.2)

O-89

Cover image: www.pixabay.com

Generis Publishing
Online orders: www.generis-publishing.com
Orders by email: info@generis-publishing.com

DEDICATION

This thesis is dedicated to my husband Stephen Otiang'a Mosi, My children: Job Otieno, Yvonne Oranga and Joash Mosi because they gave me the support I needed at various levels of my education.

ACKNOWLEDGEMENT

This thesis is the precipitation of research work spanning a period of over ten years. The realization of it would not have been possible without the assistance of many people. I therefore wish to acknowledge contributions of several people who are important to me and directly or indirectly played significant roles in successful completion of this research work. My heartfelt gratitude goes to my supervisors: Dr. Ayere, A. Mildred and Dr. Rabari, A. Joseph whose patience, continued guidance, invaluable support, insight, advice, stimulating discussions, encouragement and assistance have made it what it is.

I express my appreciation to all the Sub-County Quality Assurance and Standards Officers in charge of ICT in Kisumu County, principals, teachers and students of the pilot study and actual study schools for facilitating data collection at their stations, and some who allowed me in the back of their classes. My sincere gratitude goes to Dr. Dalmas Owino for his assistance in data analysis. I owe special gratitude to Karapul Mixed Secondary School staff for their support and encouragement during this study. Special thanks to my then, deputy principal Mr. Enos Okumu for bearing the extra load as this work was in progress. My sincere appreciation goes to all my family members especially my daughter Yvonne Oranga, sons Joash Mosi and Job Otieno Okoko for their moral, financial support and prayers.

Lastly, I acknowledge the role played by Dr. R. Ongunya of Masinde Muliro University of Science and Technology for all the assistance and igniting my interest in Information Communication Technology Integration in Education. All honor, glory and praise to the Almighty God who endowed me with health, strength, patience and perseverance during the study.

ABSTRACT

Information Communication Technology (ICT) integration is embraced in education to improve quality of learning and performance globally. In 1996, the Ministry of Education in Kenya, directed secondary schools to introduce computer studies into their curriculum. In 2006 (with studies having demonstrated positive correlation between ICT integration and performance) under its national ICT policy, the ministry encouraged ICT integration in education using even basic infrastructure including mobile phones and Internet; and, jumpstarted the program with grants for ICT equipment to 5 schools in every constituency in the republic. Chemistry is expected to benefit more because the versatile ICT tools concretize concepts and enhance learning of its abstract content. However, in Kisumu County Chemistry performance has remained the lowest among neighbouring counties, with an average mean score of 4.45 out of 12.00 from 2013 to 2016, suggesting that despite ICT integration, much has not been achieved. Moreover, training of teachers on computer use as a tool for ICT integration has been haphazard, rising doubts on teacher preparedness and their perceptions of ICT integration. The purpose of this study was to assess ICT preparedness, integration in education and stakeholders' perceptions of ICT on Chemistry performance among public secondary schools in Kisumu County. Objectives were to: establish availability of ICT resources, establish Chemistry teachers' capacity to integrate ICT in pedagogy, establish the extent of integration of ICT and assess stakeholders' perceptions of ICT integration on learner performance in Chemistry. Theoretical stimulus and response model involving Teacher Quality, Instructional Quality and Student Outcome was adopted. Descriptive survey and sequential exploratory design was used. The study population was 122 Chemistry teachers, 61 school principals, 5,962 Form 4 Chemistry students from 61 public secondary schools with computers and 7 Sub-county Quality Assurance and Standards Officers (SQASO) in Kisumu County. Stratified random sampling technique was used to sample 17 secondary schools with computers. Student sample size was determined using Krejcie and Morgan Table which yielded 340 Form 4 Chemistry students, while 39 Chemistry teachers and 17 school principals were purposively selected. Saturated sampling was used to select 6 SQASOs. Data collection instruments were Questionnaires, Observation Schedule and Lesson Observation Checklist. The instruments were piloted on 10% of population of principals, teachers, students and SCQASO. The Chronbach's alpha reliability indices (α) of the questionnaires for principals, Chemistry teachers and Chemistry students were .70, .72 and .71 respectively. Content validity of the instruments was established by experts' advice at the School of Education, Maseno University. Quantitative analysis of data was done using descriptive statistics that included frequency counts, percentages, means and standard deviations, while qualitative data was analyzed thematically. Findings indicate that ICT resources, only available at the ICT centers, are not adequate for integration in chemistry pedagogy; all Chemistry teachers are IT literate but seldom perform tasks in lesson preparation or in pedagogy with ICT; ICT enhanced environment does not exist in Kenyan classrooms as only 24.5% of learners have experienced some form of ICT integration in their chemistry lessons; and, stakeholders have positive perceptions that ICT integration improves performance but seldom use it in pedagogy. The study concluded that traditional methods still dominate Kenyan education system; and recommends that the Ministry of Education considers funding ICT resources in schools as it does other obligations under Free Day Secondary Education (FDSE); teachers should train on technological integration; all schools should have websites and go digital with all their pedagogical activities.

TABLE OF CONTENTS

LIST OF ABBREVIATIONS AND ACRONYMS

CD	Compact Disc
CD-ROM	Compact Disc-Read Only Memory
CEMASTEA	Center for Mathematics and Science Teachers
DEO	District Education Office
DQASO	District Quality Assurance and Standards Officer
E-learning	Electronic learning
E-school	Electronic school
E-library	Electronic library
ESP	Economic Stimulus Program
ICT	Information Communication Technology
KCPE	Kenya Certificate of Primary Education
KCSE	Kenya Certificate of Secondary Education
KESSP	Kenya Education Sector Support Program
KICD	Kenya Institute of Curriculum Development
KNEC	Kenya National Examination Council
LCD	Liquid Crystal Display
MPITE	Master Plan for IT in Education
MOEST	Ministry of Education Science and Technology
NEPAD	New Partnership for Africa's Development
OHP	Over Head Projector
TV	Television
UNESCO	United Nations Educational Scientific and Cultural Organization

CHAPTER ONE

1.1 Introduction

This chapter gives the background to the study which is then summarized into the statement of the problem. The chapter also presents the purpose of the study and the study objectives including the research questions, scope, assumptions, limitations, significance of the study and the theoretical framework. It ends with operational definitions of terms as used in the study.

1.2 Background to the Study

Globally schools use Information Communication Technology (ICT) to create, disseminate, store and manage information (Bitner & Bitner, 2002). ICT are defined as computers, Internet service provision, broadcasting technologies (radio and television), audio conferencing, telephony, network-based information services and library and documentation centers; digital resources and those that can be converted into or delivered through digital forms, specifically having relevance to education (Sundipta, 2015; UNESCO, 2002; MOE, 2012; Haddad and Drexler, 2002; Butzin, 1988). From the definitions, full potential of ICT as educational tools will remain unrealized if they remain merely for presentation or demonstration.

In some contexts around the world, ICT has become integral to the teaching-learning interaction (Bonfillus, 2019) through such approaches as replacing chalk-boards with interactive digital whiteboards, using students own smart phones, or other devices for learning during class time, and "flipped classroom" model where students watch lectures at home on computer and use classroom time for more interactive exercises. In Europe, Marilyn & Norbert, (2006) established that an appropriate integration of ICT in school education is considered a key factor in improving quality at this educational level. According to the researchers, the European commission is promoting the integration of ICT in learning process through its E-learning action plan, one of the aims of which is "to improve the quality of learning" by facilitating access to resources and services as well as remote exchange and collaboration. This would help Kenyan Ministry of Education realize its objective of digital platform with E-learning (MOE, 2012).

Further, in Asia, beyond sub-regional differences, the internal digital divide of developing countries has increased significantly as urban centers quickly adopt ICT; while it remains out of reach for rural and remote regions (Zang *et al.,* 2012). In America, Stuart (2015) explains two main ICT integration approaches from United States (US) public schools: ICT as a teaching and learning tool and ICT as a subject matter that is integrated into a variety of other subject matters (ISTE, 2000). Concluding thus, in order to prepare students with skills and knowledge necessary for the information society, ICT should be integrated to all levels and all subject matter in curriculum appropriately. By contrast, in Kenya and indeed the rest of Africa emphasis is on ICT as a clear cut subject area, taught as computer studies; the main reason why it is not considered as a learning tool in these parts of the world.

In Africa, Karsenti, Collins and Harpen (2012) investigated availability and usage of ICT resources on performance in a Pan African research Agenda. Findings indicate that only one quarter of the schools in the Pan African survey had all their computers connected to the Internet, 81.8% were urban schools compared to 50% semi-urban and rural, indicating inequalities in application of ICT to education.

Ferrell *et al.* (2007) conducted a study on NEPAD e-schools projects and found high costs as main barrier, as roughly only 10% of secondary schools with computers are able to share teaching resources via Local Area Network (LAN) and, the student-computer ratio is 150:1. Ayere (2009) researched on NEPAD and Non-NEPAD schools in Kenya by comparing ICT application areas in education and development. The findings indicated that NEPAD schools had higher quantity and better quality of ICT equipment than non-NEPAD schools that resulted in significant improvement in the performance of the NEPAD schools.

The government of Kenya through the ministry of education in the year 1996 directed all secondary schools to introduce computer studies in their curriculum (MOEST, 2005a). The current study targeted the schools that acquired computers to comply with this directive and sought to investigate if they use the equipment for ICT integration to teach all curriculum subjects, this was in an effort to establish whether the availability of ICT equipment at an institution creates a stimulating environment for learners to access and utilize.

In the year 2006 the government of Kenya developed a national ICT policy whose vision was to develop a prosperous ICT-driven Kenyan society and mission to improve the livelihood of Kenyans by ensuring availability of accessible, efficient, reliable and affordable ICT services, in its ICT policy in education, the ministry directed all schools to integrate ICT in teaching and learning all curriculum subjects (MOE, 2006). One vital step towards this was granting computers and other related medium of instruction in pilot schools, five schools in each constituency received computers to structure ICT centers.

The project never went beyond the pilot stage; but as a ministry policy, schools are expected to integrate ICT using even basic infrastructure including Internet and cell phones. Schools with computers from government grants were also targeted in the current study.

In Kisumu County, Oyoo (2018) investigated perceptions of teachers on motivational strategies on student choice of computer studies as a subject in secondary schools; and established that motivational strategies were highly effective to encourage students to opt to study computer studies; implying that in Kenya computers are feared by learners and are not considered as learning tools. Nonetheless analysis of available ICT resources for educational integration of ICT in the classroom to teach all curriculum subjects in Kisumu County is not known. This led the researcher to investigate availability of ICT resources in public schools in the county to address the gap.

Globally, digital literacy is defined as the skill of searching for, discerning and producing information as well as the critical use of new media for full participation in society (Tong & Trinidad, 2005). In many countries of the world digital literacy is being built through the incorporation of ICT in schools. It is also virtually known in all countries that the key predictor of student learning and performing well is the quality of their teachers (Olulube, 2005). Therefore an effective teacher education program is prerequisite for reliable education (Steketee, 2006; Lawal, 2003; UNESCO, 2002). As Bevernage, Cornille and Mwaniki (2005); Nolan (2004); Jones (2003); Jager and Lockman (1996) aver, teacher education providers are responsible for the future of teaching, they hold an important role in the process of educational motivation and implementation of ICTs. Educational reformers have long noted that teachers teach as they were taught. According to Afshari and Samah (2009); Java (2004) and UNESCO (2002), it wouldn't be possible to produce new generation teachers who effectively use new tools for teaching and learning unless teacher educators model effective use of technology in their own classes. This led the researcher to seek to establish teacher preparedness in integration of ICT in pedagogy to address the gap among Kenyan teachers on ICT literacy and use of the technology.

Several factors influence teacher integration of ICTs to instruction as follows: teachers' pedagogical and subject knowledge; technologies available or provided; teachers' attitude and confidence on the use of ICTs; knowledge and skills in ICT; perceptions on use and benefits of ICT; ability to integrate ICTs (Ruales & Andriano, 2011). In the United States of America (USA) Kounenou *et al.* (2015); Pelgrum and Law (2009) hypothesized in a model that higher levels of skills, knowledge and tools would produce higher levels of technology integration that would reflect on students' achievements positively. For this study, there was need to establish if skill levels automatically result in technology integration levels.

In his study Jones (2003) surveyed data collected in 2001 in metropolitan Melbourne in forty six primary school classrooms over a four week period, findings indicated very low extent of ICT use by teachers. Every classroom surveyed had access to computers, but their use was described as occasional; more than 90% of the supervising teachers used their classroom computers with students once per week or less for four week teaching practicum. Despite the numerous plans to use ICT in schools, teachers have received little training Afshari and Samah (2009). Separately, Wachiuri (2015); Afshari and Shamah (2009) hold the view that when technology is introduced into teacher education programs, the emphasis is often on teaching about the technology instead of teaching with technology, resulting in inadequate preparation, the reasons why teachers do not systematically use ICT in class. This applies to Kenyan teachers who are ICT literate from pre-service and in-service training, but hardly integrate the technology in pedagogy; it is important to investigate the cause of the missing link.

In Turkey, in 1985 to 1986 and in 1990 the curriculum for ICT in Teacher Education (ITE) program was reformed from theory-laden courses to more practice based courses (Alev, 2003); two courses are included: computer instructional technologies and material development. Usun (2009) explains that despite clear need to prepare teachers for level expertise in the United Kingdom (UK), teacher training curriculum in ICT seems to focus more on pedagogical skills of teachers, and not on skills for integration of ICT in pedagogy. This motivated the current researcher to seek to establish capacity of chemistry teachers in Kenya to integrate ICT in teaching and learning.

In the African continent, Unwin (2005) carried out a research study to explore the gulf between the rhetoric of those advocating the use of ICT in education in Africa and the reality of classroom practice, stating that fundamental principles of good ICT practice must be addressed for such programs to be effective in Africa: a shift from an emphasis on "education for ICT" to use of "ICT for education"; encompassing an integration of ICT practice within the whole curriculum.

Findings from Unwin (2005) indicate that there are existing interventions of ICTs in teacher training in Africa (Anderson & Elloumi, 2004; UNESCO, 2002b; Cabanatan, 2001; and Jenkins, Lieberg & Stieng, 1998). There is necessity for teachers to first be trained in basic ICT skills (UNESCO, 2002b), after which there is need to address four competencies: Pedagogy, collaboration and networking, social issues and technical issues. Unwin's (2005) report commented on objectives of NEPADs e-Africa program as listed by the coordinator (Kinyanjui, 2004) whose initiative aimed to connect more than half a million primary and secondary schools in Africa to the Internet; commenting that teacher training is important; but, according to Unim (2005), until the core emphasis shifts away from focus on simply getting schools connected, to deeper understanding of how this can transform children's learning experiences, it will be doomed to fail. NEPAD program never passed the pilot trial phase to full implementation in selected countries in the African continent.

In Kenya annual In-service training of all mathematics and science teachers by CEMASTEA from 2012 emphasize use of ICT in teaching mathematics and sciences as a way of enhancing delivery in the 21st century (CEMASTEA, 2018). Muriithi (2017); Mwangi and Khatete (2017) examined the teacher professional development needs for pedagogical ICT integration in Kenya focusing on lessons for transformation. Findings revealed variance in the use of ICT by experienced teachers, especially between personal use and pedagogical use. Most teachers felt that the approaches used in professional development did not equip them adequately for independent ICT usage in pedagogy.

Perhaps it is the case with Kenyan teachers who are IT literate but hardly use ICT in pedagogy; this can be established through a study. In Kisumu County, Otieno (2018) investigated the influence of teacher preparedness on integration of digital technologies in early years of education in Kenya. The findings showed a significant (n=202; r=.711; P<0.005) highly positive correlation between status of digital literacy and integration of digital technologies. The study recommended that the ministry of education should enhance their supervision on the implementation of integration of digital technologies in classroom teaching. There is inconsistent information on teachers training on ICT use in pedagogy among Kenyan teachers. There is need for clarification on basis of emphasis on use of the technology among teacher trainees; or teaching with technology in pre-service training and in-service capacity building for practicing teachers.

Globally, resource constrained contexts are considerable, and, therefore affect the extent of integration of ICT in teaching and learning thus: training of teachers and administrators, connectivity, technological support and software amongst others (Barak, 2007). Schools in some countries around the world, in order to ease burden of supply of ICT equipment, have began allowing students to bring their own mobile technology such as laptops, tablets or smart phones into class, rather than providing such tools to all students, an initiative called "Bring your own device" (Denisia & Suresh, 2013). Such initiative however, may result in inequality as some parents may not afford the same for their children; under such circumstances schools are tasked to ensure equality.

Bonfilius (2019); Sangra and Gonzales (2016); Kisirkoi (2015); MOE (2012); and Alvardo (2011) define the extent of integration of ICT resources as the degree of using any technological tools such as Internet, e-learning technologies and CD ROMs to assist teaching and learning. Several theories have been advanced to describe the extent of integration of ICT in teaching and learning and performance of learners in education. Spiro *et al.* (1992) and Kolb (1991), share the view that ICT as a source of motivation (stimulant) is a necessary component in learning, because it causes the learners' sensory apparatus to be activated (stimulant). Relevance, curiosity, fun, accomplishment, achievement, external rewards and other motivators (stimulants) facilitate ease of learning (response) in the stimulus and response theory.

In Europe a report by Morante, Lopez, Fernandez and Miranda (2014) indicated that since 1998 the European governments have invested some £ 1.8 billion in the national grid for ICT equipment and teacher training with the aim of helping teachers use ICT integration to raise standards, performance and transform teaching and learning. The average number of desktops available is about 246 per school; while the student to computer ratio was averagely 3.6 pupils; almost 88% of secondary schools are connected to the Internet and classrooms have computers and access to Internet is high; despite the wide diffusion of ICT and the high connection rates, European schools tend to use the technology primarily for presentational and display; and, use beyond basic level is fairly limited. This is relevant to the current study in Kenya where supply of computers to schools is still at formative stages, and integration is done in a context that views ICT utilization from similar perspectives of presentation and display.

In Asia, Shahid (2015) carried out a study to investigate the extent to which Asian science teachers are equipped with ICT skills and their self confidence in the use of ICT. The study randomly selected a sample of one hundred and ten trained science teachers from different countries of Asia working in Malaysia, Maldives, Sri Lanka, India and Singapore. The study identified that the extent of application skills of science teachers in Asia was significantly low when compared with training they had acquired; and though the respondents were from different cultures, as teachers, their application level had remained almost the same without any difference on gender. This finding is shared by Higgins (2010) who also expresses the same view.

In South Africa, Pandayechee (2017) carried out a survey on the extent of ICT integration in South African schools. According to the researcher, extent of ICT integration in education in South Africa had been severely limited by operational, strategic and pedagogical challenges. From the findings, the study established that the extent of uptake of technology is low in South Africa. The researcher in his conclusion described it as a misconception that merely providing technology can transform education. He cited challenges that lie not only on how to use the technology, but also on how to integrate digital technologies effectively into the curriculum. This was a motivation to the current study to investigate extent of usage of ICT in Kenyan schools.

In Eastern Africa, Abdelrahman, Howre and Osman (2013) investigated the extent of integrating ICT in teaching and learning mathematics and sciences in fifty Sudanese secondary schools in Khartoum. Objective was to ascertain the extent to which ICT is integrated with teaching programs of science (chemistry and physics). The study used questionnaires to collect data. Findings indicated that: majority of teachers in the sample schools did not integrate ICT in science and mathematics classrooms; and none of the schools used computers in teaching and learning scientific activities for their students, these observations were attributed to: lack of training courses for teachers to use ICT; limited time for planning; examination pressure; fear of not being able to complete syllabus; inadequate infrastructure; lack of Internet; absence of any kind of ICT management system in most schools and negative attitude of some teachers (Abdelrahman *et al.*, 2013) The findings were relevant; nonetheless the method of data collection by questionnaires alone was prone to respondent bias. This gap has been addressed in the current study by using lesson observation of individual teachers in their classrooms.

In Kenya, Karenji (2016) investigated integration of ICT in teaching English in secondary schools in Nyakach sub-county, Kisumu County. The problem of the study was that only a few schools have embraced the use of ICT in teaching and learning despite its enormous benefits in everyday life in and out of school. The study sought to investigate the extent to which teachers were using ICT in teaching. Findings showed that the use of ICT in teaching was still in the formative stages and faced various challenges; the available ICT resources were only occasionally used. Recommending that ICT be fully integrated in the education system, and intensive resource mobilization is put in place to enable schools acquire ICT resources. A similar study has not been done in teaching chemistry in Kisumu County.

Globally, there is wide spread perception that ICT can and will empower teachers and learners, transforming teaching and learning process from being highly teacher-dominated to student-centered, and that this transformation will result in increased learning gains for students (InfoDev 2020; Beaurain, 2016); creating and allowing for opportunities for learners to develop their creativity, problem solving abilities, informational reasoning skills and other high order thinking skills. However, according to InfoDev (2020) report, there are currently very limited unequivocally compelling data to support this; it is however believed that specific uses of ICT can have positive effects on student achievement when used appropriately to complement teachers' existing pedagogical philosophies.

Lindfors (2007) investigated the perceptions of a group of European chemistry teachers on the use of ICT in teaching. The findings revealed a perception that teaching was the most important thing and ICT as a tool or method was on the second place, and recommended a balance between ICT and traditional teaching; concluding that Teachers have a perception that use of ICT has to be combined with traditional classroom based teaching from pedagogical point of view. Universally teachers face similar challenges and share similar experiences; it would be important to establish if teachers share the same perception globally.

In Rwanda Munyengabe, Yiyi and Hitimana (2017) investigated primary teachers' perception on ICT integration for enhancing teaching and learning through the implementation of One Laptop per Child (OLPC) program in primary schools. Findings indicate that teachers are aware that ICT will help learners do their self coaching and that teachers have positive feelings towards ICT integration into teaching and learning process; concluding that the success of the program depends on teachers perceptions. The conclusion is relevant to this study where computers are only available at the ICT center, the perception of teachers to use them is important.

In Kenya, Mwendwa (2017) investigated the perception of teachers and principals on ICT integration in the primary schools in Kitui County, Kenya. The sample was drawn from three hundred and eighty eight public primary schools in the county. Data was collected using questionnaires for teachers and interview guides for the principals. The results indicated that most teachers had positive perception towards the use of ICT hardware and software tools in their instructional process, if only the required resources and facilities for ICT integration in the curriculum were present in their schools.

By knowing educational stakeholders perceptions on ICT integration, there will be an understanding of what they do with technology in relation to their work. No research has been done on perception of stakeholders on ICT integration in Kisumu County; there is need to carry out such a study to establish teachers view on innovative use of technology in teaching and learning.

Chemistry is a science subject that has been persistently performed poorly over the years, while the other subjects like biology and physics the mean scores are fairly high. It is the policy of the government of Kenya that candidates sitting for national examinations must enroll for at least two science subjects as a requirement (KNEC, 2012). For most schools therefore, the one of the compulsory sciences is chemistry. Physics and biology are however left optional for willing candidates; most candidates opt for biology as the second science. Because of this, the enrollment for chemistry and biology are almost five times as high compared to physics. In all the sciences the syllabus is examined in three papers namely Paper 1, 2, and 3. Paper 1 and 2 are theory papers, while Paper 3 is a practical paper in all the science subjects.   The maximum score for all the papers is 200 before conversion to the final grade as shown in Table 1. The performance per subject for the three science subjects for five years from 2013 to 2017 is shown in Table 1.

Table 1: Candidates Overall Performance in KCSE Chemistry, Biology and Physics from 2013-2014

Year	Maximum score	Chemistry		Biology		Physics	
		Enrollment	Score	Enrollment	Score	Enrollment	Score
2013	200	439,847	49.00	397,319	63.26	83,162	82.63
2014	200	476,582	64.31	432,977	73.65	83,162	73.42
2015	200	515,888	68.71	465,584	79.59	83,162	62.62
2016	200	566,836	47.42	509,982	68.37	109,811	70.22
2017	200	606,515	48.09	545,663	67.85	110,212	73.11

(*Source: www.kcse-online.info)

Table 1 shows the comparison of chemistry performance with biology and physics. The scores indicated out of maximum of 200 for each paper are shown per subject. Though the enrollment in chemistry is slightly higher than biology, the enrollment in the two sciences is almost five times that of physics. It is worth noting that apart from the performance of chemistry being the lowest among the three sciences, it also fluctuates erratically over the years indicating uncertainty in future outcome.

To investigate performance in chemistry, the current researcher was interested only in schools that already have computers; therefore a baseline survey was done to establish which schools have computers in the county by visiting the office of the county director in an effort to obtain the required list. A list was therefore obtained of all schools that teach computer studies and all the schools that got government grants through the Economic Stimulus Program (EPS) to procure computers and start an ICT center in their institutions; a total of 61 schools were established to have computers and have ICT centers in their institutions.

According to data on examinations analysis of KCSE from the year 2011 to 2016 at the office of the County Director of Education (CDE, 2017); the annual analysis booklet of Kisumu County Education and Prize Giving Day, the mean grade in chemistry has remained D with over 78% of 61 schools with computer laboratories in the County maintaining a mean grade of D (D+, D, D-) and below in chemistry as shown in Table 2.

Table 2: KCSE Chemistry Performance from 2011 to 2016

	A	A-	B+	B	B-	C	C	C-	D+	D	D-	E	Mean grade
2011					20	12		204	240	220	214	410	3.8 D+
2012				1		32		203	250	227	212	470	4.1 D+
2013					12	29		204	240	223	213	480	3.9 D+
2014					21	31		214	250	238	228	479	3.8 D+
2015					19	42		329	264	275	384	529	3.9 D+
2016				1	23	43		351	273	249	376	479	4.3 D+
Total				2	104	189		1505	1517	1432	1627	2847	4.0 D

(*Source CDE Kisumu)

The analysis in Table 2 indicates that among the study schools in Kisumu County out of the 61 public schools with computer laboratories a mean grade of B+ and above has never been attained for the last six year period. From the raw data, less than 7% and 14% of the 61 schools attained a mean grade of C+ and above and C and above respectively during the same period. In 2011 to 2016 a total of 78.5% of the schools attained a mean grade of D (Refer to Table 2). The education stakeholders in the county are concerned with the need to raise quality grades in the sciences, chemistry included (CDE, 2017). The county director suggested the infusion of ICT in the curriculum to make abstract concepts simpler. It is not known if the analysis in Table 2 could be attributed to inadequate ICT resources, inadequate science laboratory resources, and failure of teachers to integrate ICT in pedagogy effectively or difficulty level of the subject. Data on KCSE chemistry performance in Kisumu County and selected counties neighboring Kisumu County in the western region was obtained from KNEC and summarized as shown in Table 3.

Table 3: Comparison of KCSE Chemistry mean grades from 2013-2016 per County

County	2013	2014	2015	2016	Mean
Kisumu	4.6	4.1	4.7	4.4	4.45
Homabay	4.9	4.7	4.8	4.9	5.83
Siaya	4.3	4.6	4.7	4.8	4.6
Nandi	4.9	4.7	4.5	4.8	4.73
Vihiga	4.7	5.2	4.8	6.0	5.18

(*Source: Kenya National Examinations Council, 2017)

Table 3 indicates that the mean grade of Kisumu County is 4.45 out of 12.00. According to Cohen and Hill (2001), to perform is to produce valued results. A producer can be an individual or a group of people engaging in a collaborative effort. According to the expert, good performance means the meeting by the company 50% or more of the amount of target for any stated fiscal year. In this study the performance is measured against KNEC mean score of 12.00 reflecting grade A.

Table 3 also indicates that the mean grade for Kisumu County is the lowest among neighboring counties in the region. Improvement can be structured from the county level (CDE, 2017), Kisumu County being a region of upcoming industrial hub. There is need to raise the mean grade from 4.45 compared with other neighboring counties. Research shows that there is a significant relationship between ICT use and academic performance of learners. How teachers value integrating ICT in teaching and their perception of the technology on their learners performance need to be established.

1.3 Statement of the Problem

Worldwide ICT integration is embraced in education to enhance teaching and learning; retention and comprehension of academic content through use of images and enables interactive lessons and interesting classroom sessions which could in turn improve attitude and performance of learners. Moreover, it assists to explain complex concepts and processes through interactive media, animations and photographs thus ensuring learner understanding, quality of learning and performance. In spite of the enormous benefits of ICT in and out of school, institutions of secondary education have not embraced the use of the technology in teaching and learning curriculum subjects; even when the schools have computers.

In collaboration with UNESCO and the private sector, the government of Kenya has attempted put in place several initiatives to enhance ICT integration in secondary schools. Despite that, preparedness in terms of availability of ICT resources, teacher capacity to integrate ICT, and the extent of integrating ICT in pedagogy is largely unknown in Kenya, especially Kisumu County. To date less than 10% of secondary schools in Kenya can be said to own computers and able to offer ICT integration to teach curriculum subjects despite its perceived role in education.

The potential and efficiency of ICT equipment varies according to how it is used. The extent of integration of ICT in education should be determined by the extent it is deployed for realizing the goals of teaching and learning, enhancing access to resources, building capacities and management of educational system. By knowing educational stakeholders perceptions on ICT integration, there will be an understanding of what they do with technology in relation to their work. Performance of chemistry students in Kisumu County, Kenya, has been low for a long time. According to Kenya National Examinations Council (KNEC), the performance of chemistry was 4.45 out of 12 for four year period, which is relatively low, bordering grade D. With such grades, it is difficult for learners graduating from high schools to pursue science related courses, especially at tertiary institutions and universities. In view of this discrepancy there is need to devise chemistry pedagogy integrated with ICT to make its abstract concepts easier for secondary school students to understand and improve their performance. If this is not addressed, poor performance will discourage secondary school learners from pursuing the subject, and, when made compulsory as it is presently the case, the performance will continue to dwindle at the expense of the learners in this country.

1.4 Purpose of the Study

The purpose of this study was to assess ICT preparedness, integration in education and stakeholders perceptions on chemistry performance among public secondary schools in Kisumu County, Kenya.

1.5 Objectives of the Study

The objectives of this study were to:

1. Assess availability of ICT resources for teaching Chemistry in public secondary schools in Kisumu County.
2. Establish Chemistry teachers' capacity to integrate ICT in pedagogy.
3. Establish the extent of integration of ICT in teaching and learning Chemistry in public secondary schools in Kisumu County.
4. Assess Stakeholders' perceptions on the influence of ICT integration on KCSE Chemistry performance.

1.6 Research Questions

The following research questions were used to guide this study:
1. What is the level of availability of ICT resources for teaching Chemistry in public secondary schools in Kisumu County?
2. What is the capacity of Chemistry teachers to integrate ICT in pedagogy?
3. What is the extent of ICT integration in the teaching and learning of Chemistry in public secondary schools in Kisumu County?
4. What are the Stakeholders' perceptions on the influence of ICT integration on KCSE chemistry performance?

1.7 Scope of the Study

This study was conducted in Kisumu County and focused on 61 secondary schools with computers. These included schools that offer computer studies as a subject at KCSE and therefore had computer laboratories; and, Economic Stimulus Program (ESP) secondary schools that benefitted from the Economic Stimulus in the year 2009 by getting grants from the government to start ICT centers with computers in their institutions; this gave a scope of 61 secondary schools with computers in Kisumu County for the study. Though only 61 secondary schools in Kisumu County were covered out of more than 250 secondary schools in the county, the types of schools targeted were based on national criteria. The schools are distributed in the seven sub-counties that are also politically referred to as constituencies in the government of Kenya programs. The Sub-Counties in Kisumu County are: Kisumu East, Kisumu West, Kisumu Central, Seme, Nyando, Muhoroni and Nyakach.

The study involved school principals, chemistry teachers and Form Four chemistry students and SCQASO as the stakeholders. School principals were selected for this study because as administrators, it is their duty to identify and implement strategies that can improve educational outcomes of their students. Chemistry teachers were selected on the basis that they were teaching the Form Four chemistry class in the year of the study; such that their responses would be compared to those of the Form Four chemistry students' respondents.

The Form Four chemistry class was chosen for the study because according to Peters (2010), Kennedy and Cox (2008) they are positively connected to their peers, other students in school and their teachers; furthermore, their teachers also know them including their strengths, interests and learning needs. They were therefore considered the most appropriate class to involve in the study. Moreover, their KCSE performance in chemistry formed part of data utilized for analysis. The study also targeted Sub-county Quality Assurance and Standards Officers (SCQASO), one from each sub-county who are in charge of all ICT programs in their respective sub-counties. These are the informed experts in ICT who formed the specialist category of respondents.

1.8 Assumptions of the Study

The following assumptions guided the study:

1. That the Ministry of Education and the Teachers' Service Commission ICT policies requiring schools to teach all subjects integrated with ICT applies.

2. Schools equipped with computers are expected use them for teaching, not just computer studies as a subject, but also ICT integrated for teaching and learning other curriculum subjects offered in the public schools.

1.9 Limitations of the Study

The major limitations of this study are:

1. Use of questionnaires to obtain data for analysis. This research relied on the use of questionnaires to obtain data for analysis. These could have had limitations arising from respondent bias, a partiality that prevents objective consideration of an issue or situation. This was

minimized by eliciting the same facts from different people in the same setting: the school principal, the chemistry teachers and the chemistry students. This is triangulation, done in order to increase reliability of the instruments as explained by Ferrel *et al.* (2007). This limitation was further minimized by use of observation schedule that assisted the researcher with first-hand information on the location and actual usage of ICT equipment in each school; and a Lesson observation checklist that was used during lesson observation of ICT integrated lessons.

2. Type of schools from which data was collected. In Kenya, public schools exist as national, extra-county, county or sub-county hence schools are of different types. The study schools in this research were of different types and hence were intervening variables. It is important to note that however, the school type was not a factor of analysis in the study and were considered as an intervening variable because KCSE results used in the study were not due to ICT integration level alone, but could have actually been due to school type. This limitation was minimized by the fact that the KCSE results were not used to rank the schools (the school names were given numerical values without any order) but was used as reference to determine if integration of ICT took place or not in the schools; what impact would it have on the given performance. Therefore, intervening variables that include school type, culture, environment and infrastructure of a school were treated as controls that were not expected to impact on the results of this study.

3. Principals, teachers and students of the study schools who participated in the study attributed their responses on their experiences as at the period 2017/2018. However the KCSE results used in the analysis covered the period 2012 to 2017 hence cannot entirely be attributed to the independent variables in the study as the situation has not been constant. It was important to highlight this as any discrepancy that would arise was minimized by calculating the mean score for the stated period that was used in data analysis; therefore this would not have an adverse effect on the final analysis as the trend on performance of a school is based on the culture of the school, while administrators, teachers and learners come and go.

1.10 Significance of the Study

The findings of this study could influence the development of appropriate and relevant guidelines by the curriculum developers on the usage and uptake of ICT by teachers, starting from teacher training to practicing teachers in schools by involving all stakeholders in the program. The study could further be significant to the ministry of education to enable informed decisions on enhancement of ICT use in education, especially on the emphasis of e-learning school work at home during lockdown when schools closed for a long period due to Covid-19 pandemic.

1.11 Theoretical Framework

This study was modeled on the theory of Stimulus and Response model advanced by Nilsen, Gustafsson and Blomeke (2016). This model suggests that Teacher Quality and Instructional Quality determine Student Outcome. The model postulates that "stimulus elicits response", and specifically defined these relationships; the variable "teaching methods by integrating ICT" represented the stimulus, while the "students' academic performance" represented the response. The theory further explains that "Stimulus" deals with the communications the learner receives; and "Response" focuses on the learner to produce the responses that are called for by the communication. Besides, the theory hypothesizes a direct link between amount of practice, massing and distribution of trials, the schedules of reinforcement, and other variables acting as stimuli that strengthen the learner's response to take place.

The theory was used by Klieme, Pauli, and Reuser (2009), Seidal and Shavelson (2007) and established that the relationships have been difficult to quantify and understand empirically. The research pointed to challenges in measuring teacher and instructional quality. The theory postulates that a relationship exists between teacher quality, instructional quality and learner achievement. This theory was preferred over Creemers and Kyriakides (2008) and Chapman *et al.* (2012) whose limitations included models which could only particularly account for the nested nature of data.

This study focused on chemistry teachers' ICT literacy and perceptions of the teacher towards use of the available technology as the stimuli that guided the study. Teachers are more likely to use available ICT resources as stimuli if they perceive it to have a positive impact on their learners' performance. Chemistry teachers' perceptions towards use of ICT in teaching their subject will determine their behavioral intention to integrate the technology as stimuli in teaching and learning process.

For this study therefore, it was theorized that chemistry teachers' ICT literacy and proficiency in ICT integration in chemistry pedagogy in public secondary schools could be enhanced if teachers have positive attitudes towards ICT use of available resources, and, if they believe that ICT integration could lead to improved performance. From the theory, teachers would integrate ICT in teaching and learning and use it as a "stimulus" if they perceived that it would elicit a "response" by improving quality of education offered, hence improved performance of their learners.

As applied to this study, the theory holds that teacher instructional qualities that include professional qualifications, ICT literacy and proficiency, perceptions and extent of ICT integration influence student achievement. This is true, considering that teacher instructional quality is a construct that reflects those features of teachers' instructional practices known to be positively related to student outcomes, both cognitive and affective ones (Decristan *et al.* 2015; Fauth *et al.* 2014; Good *et al.* 2009; Hattie, 2009; Klusmann *et al.* 2008; Siedel and Shavelson, 2007). Thus, if the study on relationship of student to the quality of their teachers and instructional quality believes that teacher professional qualification, confidence, job experience and continuous professional development affect student achievement, then teacher quality has been shown to be of importance for student achievement. However, in adopting this theory, the researcher is not ignorant of substantial research gap that exists with respect to the challenges in measuring teacher and instructional quality; and, with respect to comparative research across countries applying the same kind of instruments.

The theoretical model was based on the premise that teacher ICT capacity, their perceptions, extent of integration of ICT in teaching and learning plays significant role in stimulating response by the achievement of the dependent variables.   With this in mind the researcher refined goals of the research, selected appropriate variables and methodologies and identified potential validity threats.

1.12 Definition of Terms/ Operational Definition of terms

Audiovisual - Are materials for using sound and sight to present information. In this study it includes means of teachers processing sound and visual components using both sight and sound, typically in the form of slides or video and recorded speech or music in teaching and learning.

Computer Lab- It is a space which provides computer services to a defined community. In this study the term was used to refer to a room set aside by a school and fitted with computers to provide computer services to teachers and students of that institution. In this study the computer lab is also referred to as an ICT center.

Education-In this study it comprises teaching and learning of chemistry

ESP Champions-In this study it refers to classroom teachers who were initially trained by the ministry of Education in Kenya to initiate the training of classroom teachers country wide in the integration of ICT.

ICT-Information and Communication Technology is a diverse set of technological tools and resources that consist of hardware, software, network and media for collection, storage, processing, transmission and presentation of Information to users. In this study it was used to include ICT resources and ICT literacy of teachers and levels of ICT integration.

ICT Integration- In this study it was used to refer to the act of incorporating or merging ICT in the teaching and learning of chemistry.

ICT- Integration level: In this study it refers to the level of availability and use of ICT resources, frequency of use of the ICT resources, software and devises by teachers in teaching variety of topics in chemistry, literacy and proficiency of teachers to use those devices

ICT-literacy-It is the level of training in computers. In this study it was used to include training in ICT and using digital technology, communication tools and networks to access manage, integrate, evaluate and create information in order to use the technology in pedagogy.

ICT Preparedness: It refers to the readiness of schools for ICT integration in terms of availability of ICT resources and training of teachers to build their capacity on literacy and proficiency in the use of ICT.

ICT Resources: Refer to computers, telecommunication and audiovisual systems used during ICT integration. In this study it was used to refer to, not just the presence of, but also functionality and serviceability of the equipment. It also refers to availability of the equipment for ICT integration in teaching and learning school subjects.

Intervening Variable –in this study are variables which are not independent variables but are occurring between the variables and may influence the dependent variable and could affect the results of the study. The researcher wants to make sure that it is the independent variable that has an effect on the dependent variable.

Performance: It is the recognition of accomplishment especially in exams. In this study it included accomplishment in KCSE examinations by learners as a measure of being taught by teachers proficient in performing teaching tasks (pedagogy) successfully while incorporating ICT in chemistry pedagogy. It also refers to performance of chemistry teachers in using ICT in pedagogy

Software programs: Refers to any set of instructions that directs a computer's processor to perform specific operations. In this study it includes computer program e-libraries, associated documentation and the application software stored in the computer.

Stakeholders: In this study it comprises of teachers and students of chemistry, school Principals, and SCQASO in Kisumu County

Teacher ICT-proficiency: In this study it refers to the skillfulness in the command of fundamentals of ICT derived from practice by teachers, familiarity and technique.

Telecommunication: Are systems used in transmitting messages over a distance electronically. In this study it includes the science and technology of the communication of messages over a distance using electric, electronic or electromagnetic impulses which occur when the exchange of information between a teacher and a student includes the use of technology.

CHAPTER TWO
LITERATURE REVIEW

2.1 Overview

This chapter discussed the literature related to Information Communication Technology (ICT) preparedness, integration in education and stakeholders perspectives on KCSE chemistry performance among public secondary schools in Kisumu County, Kenya. The chapter particularly focused on availability of ICT resources in public secondary schools, teacher literacy in the integration of ICT in pedagogy, extent of integration of the ICT in chemistry, and the perspectives of stake holders on the influence of ICT integration on KCSE chemistry performance. These were considered the pillars of the study.

2.2 Availability of ICT Resources in Public Secondary Schools

Globally schools use Information Communication Technology (ICT) to create, disseminate, store and manage information (Bitner & Bitner, 2002). In some contexts around the world, ICT has become integral to the teaching-learning interaction (Bonfillus, 2019) through such approaches as replacing chalk-boards with interactive digital whiteboards, using students own smart phones, or other devices for learning during class time, and "flipped classroom" model where students watch lectures at home on computer and use classroom time for more interactive exercises.

Sundipta (2015); MOE (2012); (UNESCO, 2002); Haddad and Drexler (2002); Butzin (1988), define ICT resources as diverse set of technological tools and resources used to communicate and to create, disseminate, store and manage information; digital and those that can be converted into or delivered through digital forms. The resources include computers, Internet service provision, broadcasting technologies (radio and television), audio conferencing, telephony, network-based information services and library and documentation centers; specifically communication technologies having relevance to education, and, the potential of each resource varies according to how it is used. According to Sundipta (2015) at least five levels of technology use in education are identified: presentation, demonstration, drill and practice, interaction and collaboration. From these definitions full potential of ICT as educational tools will remain unrealized if they remain merely for presentation or demonstration Globally, ICT integration is viewed as a "major tool for building knowledge society" (UNESCO, 2008) and particularly as a mechanism at the school education level that could provide a way to rethink and redesign the educational systems' processes, thus leading to quality education for all. In Europe, an appropriate integration of ICT in school education is considered a key factor in improving quality at this educational level. The European commission is promoting the integration of ICT in learning process through its E-learning action plan, one of the aims of which is "to improve the quality of learning" by facilitating access to resources and services as well as remote exchange and collaboration (Marilyn & Norbert, 2006).

UNESCO publication (UNESCO, 2008) outlines several aspects to be observed and promoted such as: widespread access to broadband technologies, professional development, and support policies for teachers, more research into how people teach and learn using ICT, development of new high quality online content and adaptation of current regulations to make the integration of ICT at schools easier. It is worth noting here that the efforts of different governments and administrations worldwide have been focusing on providing schools with good equipment. However, analysis of educational use of ICT in the classroom to teach specific subjects like chemistry is insufficient.

Asia is the largest worlds' region by land and is also the largest by population; some member states in Asia have matched or outperformed international standards in the field of ICT assisted instruction (Pedro, 2012). While in other countries upgrades to national networks, telecommunication improvements, enhanced national connectivity and the introduction of new Internet Provider (IP); delivery technologies are creating a more favorable environment for the uptake of ICT. In several least developed countries Internet based forms of teaching and learning and the essential infrastructure to support it are limited except for privileged few, driving countries to consider other forms of ICT. A similar situation may be applicable to Africa, which is also a vast continent that is economically challenged. In Asia, beyond sub-regional differences, the internal digital divide of developing countries has also increased significantly as urban centers quickly adopt ICT; while it remains out of reach for rural and remote regions (Zang *et al.,* 2012). This is relevant to the current research carried out in Africa, where digital divide arise between urban and rural schools; rural schools are highly disadvantaged in terms of access to ICT resources.

In America, Stuart (2015) carried out a study to examine ICT integration into United States (US) public school curriculum and proposed models related to middle and high school course mathematics and science subjects. The researcher explained two main ICT integration approaches from US public schools: ICT as a teaching and learning tool and ICT as a subject matter (one of content area) that is integrated into a variety of other subject matters, such as mathematics, science, literacy and technology together and as a mixed curriculum. The researchers aver ICT is a teaching and learning tool that can provide a number of learning activities from tutoring to simulations in subjects like chemistry; and, ICT is a mediation tool, a cultural artifact for the socio-cultural approaches to learning (ISTE, 2000). On American ICT integration, Stuart (2015) concluded thus, in order to prepare students with skills and knowledge necessary for the information society, ICT should be integrated to all levels and all subject matter in curriculum appropriately. This is relevant to the present study in Africa, where ICT as a subject matter is predominant and is taught as computer studies, while ICT as a teaching and learning tool is given a back stage, and, without institutional policies in place, it depends on concerned teachers' perceptions.

Becta (2005) in a research in the United Kingdom reported that when ICT are used in pedagogy, they add value to the process; and have the capacity to support teaching, learning and classroom management as efficiency is increased. The researcher reported on the evidence on the progress of ICT in education in the United Kingdom. The study sought to suggest directions for the development of new content for e-learning programs. Their main objectives were to survey the impact of ICT in education and its progress in classroom instruction for subjects like chemistry and to explore the inherent and current challenges of fully integrating ICT in education in a dynamic policy environment. The researchers concluded that there is real evidence of positive impact of ICT use in education. This motivated to the current study to seek availability of resources and influence of ICT on performance of learners.

Abbit (2011), Mishra, Harris and Koehler (2006) investigated two types of educational software resources available in Europe: Content Free Software and Content Rich Software. According to the researchers, the former provides teachers and students with tools to create their own content, dominated by open-ended software products that support creativity of the user such as word processing and graphic programs. The researchers recommended this type of software because creativity leads to critical thinking, a virtue important in developing minds. According to the researchers, word processing or desktop publishing citing Sibelius is software ideal in education for writing scores, image editing software, exemplified by paint shop which is ideal for editing digital photographs. Others include concept mapping software ideal for brainstorming and essay planning (Abbit, 2011). Such applications are relevant in education; however the studies fail to address how such software programs can be applied in secondary school classrooms situations for integration, revision and other applications that can impact on performance in subjects like chemistry.

Mishra *et al.* (2006) investigated content rich software, it accounts for a large proportion of commercially produced educational software. The software comprises multimedia content like graphics, video, sound, animation and many more, presented in structured way ranging from teaching basic number concepts to explaining complete chemical equations. The researchers explain tasks that can be supported by this type of software which in science include Explore Electrical Concepts; an example of software that consist of science simulation program exemplified by Exploring Science.

UNESCO (2014), Chai, Ho and Tsai (2012) and Mishra, *et al.* (2006) hold a similar view that programs in content rich software tend to restrict level of user control especially tutorial drill and practice skills for students; yet a professional teacher is well versed with difficulty level of topics; and can use ICT to explain complex concepts through videos, animations, photographs and more, thus ensuring learners' comprehension. The propositions overlook the important guidelines to the right content which is also useful ingredient in preparedness to ICT integration. The reports above are relevant and binding; however they fail to address how software can be integrated effectively to teach sciences in secondary schools to improve performance in chemistry and unravel its abstract concepts and nature that appear difficult to learners.

According to the later studies Duck Builder is free modeling software important in structure and bonding, while Teacher's Report Assistant lets the teacher collect, download and connect banks which can be used in any word processor. Spectrophotometer is used in spectroscopic experiments to enable read color changes at endpoint during chemical titrations. Enginda (2012), Guzey and Roehrig (2009) conducted similar studies on software use in educational evaluation and highlighted examples like hot potatoes software that creates multiple choice questions, short answer questions, jumbled sentences, crossword, matching and ordering and gap fill exercises; some of these tasks are also done by Easy Test Maker. The studies both support evaluation through testing; nonetheless, for easy access, when tasks are organized topically teachers can utilize them at specific intervals depending on the level of content coverage and this can boost understanding of concepts and impact on performance.

In Africa Abdelkarim *et al.* (2012) investigated the impact of integrating ICT on performance in schools in Morocco. In this research, ICT based education was examined through an integrated project to experimentally determine how such technology could influence the innovation and performance of students in science (chemistry-physics, mathematics and Life-Earth sciences) in Morocco. The experiment was deliberately run at middle school level (age 12-14) as it constitutes the best stage in students life to influence their choice for field of study. Two schools were selected based on their location, one in a semi-rural area of Ifrane, and another in a large city of Fes. The study focused on physical sciences (chemistry-physics) as they are taught together by the same teacher and represent a structural template of what was carried out within the framework of the pilot project.

The project was the first of its kind in Morocco since it permitted decent usage of ICT in classroom sitting and allowed integral participation of middle school teachers in the elaboration of ICT pedagogical teaching resources. Control (standard) and experimental (ICT-based class) were both taught by the same teachers and all experiments were carried out in a close collaboration with the authors and the teachers, and assistance from ICT engineers and technicians. Quantitative evaluation of the experimental data based on general balanced 3-stage nested design, together with qualitative assessment show a positive impact on the performance and motivation of students and teachers. The findings are relevant to the current study, however, considering the sample size of only two schools; the findings cannot easily be generalized to a larger population, which weakens external validity of the study. It is also worth noting that the technique of using the same teacher to teach both control and experimental class could create issues of subjectivity which could have impacted on the findings.

Karsenti, *et al.* (2012) investigated availability and usage of ICT resources on performance in   a Pan African research Agenda aimed at understanding how pedagogical integration of ICT can improve the quality of teaching and learning in Africa. The researchers' posited difficulty in the effectiveness of ICT into the learning process as statistics indicated more than one hundred students for every computer in most public schools. The success and challenges in three hundred and fifty plus public schools investigated using a sample of one hundred public schools south of Sahara were identified as Internet connectivity; none of the computers at the schools were connected to the Internet. The students were taught basic office applications using CDs to supplement library resources. According to the findings, even the highest ranked schools had neither a website nor an e-mail address; most of such schools were in the republic of Congo followed by Central African Republic where seven out of ten public schools had no Internet connections. Poorer still was Kenya, categorized together with Uganda where eight out of ten public schools surveyed lacked Internet connections. In Kenya, many public schools perform online functions required by the ministry at cyber cafes, creating a void on Internet resources Mozambique had the highest number of public schools with Internet connections, credited to existence of national ICT policy structured in the year 2000 with a focus in education, and an effective National commission for ICT policy, which focuses on strategy for implementation; by the year 2002 it had adopted School-Net as one of its flag ship projects. One quarter of the schools in the Pan African survey had all their computers connected to the Internet, though 81.8% were in the urban schools compared to 50% for semi-urban and rural. The researchers concluded that there were substantial inequalities in a number of social factors in the application of ICT to education (Karsenti *et al.*, 2012).

Though the researchers investigated pedagogical integration of ICT in a population of over one million learners in public schools, which was considerably massive, they used a sample of only one hundred public schools and focused on lack of Internet as their main objective. Moreover, their sampling procedure was flawed for such a vast continent; it could have been characterized by a larger sample to raise external validity. The researchers failed to describe major weaknesses apart from availability of Internet; because they did not go further to determine the availability and usage of a variety of software programs in ICT-Teaching and Learning processes in the schools. In view of the issues raised, the former study relates to the current on the availability and use of ICT resources and performance, but expanded to include all programs that drive the operations of electronic gadgets used in teaching and learning.

In the Kenyan situation, a research by Katitia (2012) found out that the nature of culturally specific traditions have characterized the teaching and learning practices, where teachers often view their role as "provider of knowledge" and regard their students as "empty vessels" to be filled. The researcher further highlighted the issue of content development, the lack of culturally appropriate educational content possesses challenges and recommends that content produced in one context cannot be adopted without modification into another context. The researcher concluded that for performance to improve in chemistry, ICT can be used to make the subject learner centered by involving them in search for information to solve problems. However, the study failed to establish teacher role in ICT integration by engaging learners in knowledge search through relevant software content and problem solving learner centered approach.

Achievement of this objective is still farfetched as research estimates access rate as approximately one computer to 150 students in Kenya (MOE, 2006); so far most public schools are ill equipped with ICT equipment. However, the former researcher overlooked the fact that for new technology to actually contribute to learning, more thought need to be put into neglected issues of pedagogy, curriculum, professional development of teachers, software maintenance scheduling among other issues.

The researcher appreciate the training programs conducted countrywide by CEMASTEA that have attempted to bridge the gap of digital divide and seeks to assess the extent of change impacted by CEMASTEA in-service training of teachers. Nonetheless, at the ground there is lack of follow up, monitoring and evaluation by CEMASTEA after training trends; especially how much of content learnt during in-service training is implemented by teachers concerned.

Ferrell *et al.* (2007) conducted a study on NEPAD e-schools demonstration projects and found that the core problem is Kenya lacks adequate connectivity and network infrastructure; although small numbers of schools have direct access to high speed connectivity through an Internet service provider, generally there is limited penetration of the national physical telecommunication infrastructure to rural and low income areas. From their findings they asserted that consequently there is limited access to dedicated phone lines and high speed connectivity for e-mail and the Internet. Even where access to high speed connectivity is possible, high costs remain a barrier to access. According to them, very few schools can afford to use Very Small Aperture Terminal (VSAT) technology; they estimate that roughly only 10% of secondary schools with computers are able to share teaching resources via Local Area Network (LAN). According to the document, even in schools that do have computers, the student-computer ratio is 150:1. This is not comparable to the situation in Britain where the learner-machine ratio was 8:1 in primary and 5:1 in secondary schools in the year 2006 (ISTE, 2006). In response, in an effort to provide solution to these problems, the ministry hopes to leverage the e-government initiative of networking public institutions countrywide to facilitate connectivity for the education sector (MOE, 2012).

The genesis of ICT integration in Kenya started with the government collaborating with United Nations Development Program (UNDP) in partnership with the United Nations Educational Scientific and Cultural Organization (UNESCO) and jointly launched a National Information & Communication Policy and Framework to enhance the capacity of the national information and communication sector (Tuju, 2004). The aim was to propose a national framework to improve information and communication infrastructure and to play an effective role in development. However a national framework should mirror educational framework to be effective. The government through the ministry of education in the year 1996 declared that all secondary schools should introduce computer studies in their curriculum (MOEST, 2005a), but did not explain to the schools how they were to acquire those computers (Ayere, 2009). As a result most schools failed to comply (Odera, 2002). Certain schools gradually acquired computers though government grants or independently and introduced computer studies as an examinable subject into their curriculum. It is such schools that have computers and teach computer studies as a subject that the current study sought to investigate if the computers are also used as resource for ICT integration to teach other subjects like chemistry.

In the year 2006 the government developed a national Information Communication Technology (ICT) policy whose vision was to develop a prosperous ICT-driven Kenyan society and mission to improve the livelihood of Kenyans by ensuring availability of accessible, efficient, reliable and affordable ICT services (MOE, 2006). One vital step towards this was introducing computers and other related medium of instruction in pilot schools, centers of excellence and national schools under the ESP program in order to equip the future labor force with skills to participate competitively in the 21st century education and attain Kenya's Educational Goals (MOE, 2012). This category of schools was also targeted by the current study to establish if the computers available are used to teach curriculum subjects as intended.

The national ICT policy identified four guiding principles namely: infrastructure development, human resource capacity building, co-operation between stake holders and appropriate policy and regulation framework (ROK, 2012). The policy seeks to use education as a platform to facilitate sustainable economic growth and development, wealth creation and poverty eradication, address development gaps as they relate to women, youth, rural and other disadvantaged groups, achieve progress towards full socioeconomic inclusion of all citizens through provision of universal access, stimulate investment and innovation in the ICT sector through research and development and provide for increased access to ICT services (ROK, 2009).

The research by Ayere (2009) compared the number of ICT computer accessories in NEPAD and Non-NEPAD schools. The researcher sought to investigate the input of ICT in NEPAD and non-NEPAD schools by comparing ICT application areas in education and development. Her main objectives relevant to this study were to identify quantity, quality and variety of ICT equipment and material resource available in the two categories of schools, identify significant ICT application areas and factors that influence application areas.

The methodology used was exploratory approach using descriptive survey and ex-post facto design. The findings indicated that NEPAD schools had higher quantity and better quality of ICT equipment than none NEPAD schools. The researcher recommended that institutions should make computer laboratory available to all learners just like a library is; institutional Internet and e-libraries would enhance and improve personal information base. Conclusions were that the presence of wider variety and higher quality of ICT equipment in NEPAD schools resulted in significant improvement in the performance of a school; because of the findings of a positive correlation between variety and quality of equipment to performance in KCSE.

The NEPAD program ended in the year 2009 with the government of Kenya coming up with the ICT concept as a national wide program of equipping pilot public schools with ICT equipment; however the initiative did not go further than the pilot trials (MOE, 2012). The findings in the former study indicate success in the private sector initiative to pilot programs that require funding and failure by the government to initiate full implementation to the whole republic; therefore the impact of NEPAD program has only been felt in the pilot schools. This creates inconsistency and inequality in government owned public schools.

In Kisumu County, Okewa (2011) sought to determine the types of ICTs adopted by public secondary schools, factors affecting the adoption and the benefits derived. The study adopted stratified sampling design with respondents drawn from 93 public secondary schools in Kisumu County; data was collected by use of structured questionnaires. The study found out that over 50% of respondents indicated to have adopted ICT in their schools; but more needed to be done to enhance the status of ICT adoption in public secondary schools in Kisumu County and nationally. The study concluded that: lack of ICT implementation plan, lack of finances, lack of technicians and ICT training programs remain the main factors affecting ICT adoption in public secondary schools in Kenya. The findings are relevant and binding to the current study seeking to establish availability of ICT resources in teaching chemistry; the conclusions are valid. Nonetheless, the use of questionnaires alone to collect data is prone to respondent bias and true picture of the events might not be revealed.

The cause of the low means score in chemistry is its abstract nature and difficulty level; some of the difficult concepts could be made to appear simple with integration of ICT in teaching and learning. Moreover, ICT integration when effectively done translates into improved performance (Ayere, 2009), which has not been realized in Kisumu County. If ICT equipment is available in the school computer laboratory, it is important to establish if it is sufficient or accessible enough to be applied in teaching other subjects like chemistry apart from computer studies that is examinable at KCSE. Unless an investigation is done, it is also not clear if the equipment requires to be networked inside the classrooms where teachers teach for convenience and effective integration in pedagogy. These are pertinent questions this study sought to address.

2.3 Teachers Capacity to Integrate ICT in Pedagogy

Globally, ICT literacy is defined as the skill of searching for, discerning and producing information as well as the critical use of new media for full participation in society (Tong & Trinidad, 2005). It is virtually known in all countries that the key predictor of student learning and performing well is the quality of their teachers (Olulube, 2005). Therefore an effective teacher education program is a prerequisite for a reliable education (Steketee, 2006), Lawal (2003); UNESCO (2002). As Bevernage, Cornille and Mwaniki (2005); Nolan (2004); Jones (2003); and Jager and Lockman (1996) aver, teacher education providers are responsible for the future of teaching, they hold an important role in the process of educational motivation and implementation of ICTs; for they have to anticipate new developments and prepare prospective teachers for their future roles. Teacher educators have great influence on their students. Educational reformers have long noted that teachers teach as they were taught (International Society for Technology in Education [ISTE], 2000). According to Java (2004); UNESCO (2002); it wouldn't be possible to produce new generation teachers who effectively use new tools for teaching and learning unless teacher educators model effective use of technology in their own classes. According to Afshari and Samah (2009) teachers only give to pupils what they have acquired through their training. These definitions agree on the importance of comprehensive technology based teacher training. However they ignore the context, which is also a useful ingredient of teacher training.

The potential benefits of ICT integration in teaching and learning in schools have been extensively discussed in academic literature globally. Studies show that ICT has not fully been adopted in teaching and learning environments in most schools. Ruales and Adriano (2011) argue that installation of technological devises and infrastructure will not automatically lead to integration of ICT unless teacher factor is addressed; that largely influences ICT integration in schools. Several factors influence teacher integration of ICTs to instruction as follows: teachers pedagogical and subject knowledge; technologies available or provided; teachers attitude and confidence on the use of ICTs; teachers' knowledge and skills in ICT; conceptions on the use and benefits of ICT; ability to integrate ICTs; curriculum; school leadership and support; technical support and maintenance; funds for operations; prevalent school culture; incentives and time (Ruales & Andriano, 2011).

Several models attempt to explain the levels of skills and technology integration in the classroom. According to Pelgrum (2001) the success of educational innovations depends largely on skills and knowledge of teachers. The researcher established teachers' lack of knowledge and skills was the one of most inhibiting obstacle to the use of ICTs in schools. In the United States of America (USA) Kounenou (2015) hypothesized in his model that higher levels of skills, knowledge and tools would produce higher levels of technology integration that would reflect on students achievements positively. Technological skills and knowledge are therefore important ingredients to ICT integration in schools. It would be helpful to establish if technological skills produce higher levels of integration among Kenyan teachers.

Findings from analysis of survey data collected in 2001 in a study in metropolitan Melbourne in forty six primary school classrooms over a four week period, indicated very low level of ICT use (Jones, 2003). Every classroom surveyed had access to computers, but their use could be best described as being occasional. The data collected indicated that more than 90% of the supervising teachers in the survey used their classroom computers with students once per week or less for four week teaching practicum. This is relevant to the current study; point of contrast is in Kenyan schools, there are no computers in the classrooms and classes have to move to ICT centers to access computers for integration to take place, it is therefore a wonder if integration of ICT in teaching all curriculum areas is possible under such circumstances.

Yamamoto and Yamaguchi (2016) carried out a qualitative study in Mongolia, Asia that looked at factors facilitating teachers' ICT skills, teacher morals and perceived student learning in technology using classrooms and performance. Their findings revealed that professional development has a significant influence on how ICT is embraced in the classroom. Despite the numerous plans to use technology in schools, teachers have received little training in this area through their teacher education programs (Wachiuri, 2015). According to Afshari and Samah (2009) who investigated factors affecting ICT use in Europe, when technology is introduced into teacher education programs, the emphasis is often on teaching about the technology instead of teaching with technology. The researchers concluded that this resulted in inadequate preparation to use the technology, hence one of the reasons why teachers do not systematically use ICT in class; perhaps after ICT skills have been acquired, pedagogical practice of use in policy governed institutions is required to establish proficiency of use.

Usun (2009) carried out a research to investigate ICT in teacher education (ITE) programs in the developed world and Turkey. The main problem of the study was to review comparatively, the application of ICT for teacher training. The objective of the study was to comparatively review the strategies of preparing teachers to use new ICTs such as computer assisted education and distance education in teacher education programs. The study used descriptive design through literature review; perspectives from different countries: Europe, USA and parts of Asia were examined and compared with Turkish ICT programs for teacher training. This comparative literature was relevant to the current study to give a universal view of importance of teacher ICT literacy and training on integration of the technology in pedagogy.

Usun's (2009) comparative literature indicates that many countries in Europe have official recommendations for ICT related skills for future and practicing teachers. In over half of all European countries, ICT became a compulsory part of the curriculum for the initial training of teachers for either primary or secondary education. However the official recommendations on the subject of ICT training are often general and its organizational content and the amount of time to be devoted to it are in some countries is the prerogative of the individual teacher training institutions (Balcon, 2003). Perhaps a national guiding policy on teacher training can bring harmony in the field of teacher training for the benefit of learners; this can be done with unifying recommendations from UNESCO.

In Germany, education in the teaching of ICT is one of the co-curriculum options. Consequently, the institutions of teacher education concerned are obliged to offer the subject, but not to include it in their overall course of education. This applies to the initial education of primary and secondary school teachers (Eurydice, 2004). Today the UK is the only country in most developed countries, which includes IT in its national curriculum, and the only country to have at least one computer per class in all its primary schools;

and having the best pupil to computer ratio in secondary schools. However, the researcher avers, teacher professional development in the use of ICT in teaching and learning is surprisingly still low. The finding is relevant as teacher professional ICT development is a determinant in integration of the technology.

In Belgium and Norway, courses in ICT are compulsory, but their content is an integral part of other subjects (Eurydice, 2001). In lower secondary the use of word processing and data processing software are recommended most frequently. But recommendations less frequently emphasize the command of skills such as use of educational software and Internet. Skills to integrate the technology need emphasis.

Turkey, in 1985 to 1986 and in 1990 began two distinct teacher training programs; the first was a pre-bachelor certificate for 130,000 primary school teachers, the second offered a university degree to 54,000 secondary school teachers (Demiray, 1990) as part of the National Education Development Project (NEDP) which was sponsored by Turkish government and the World Bank. Education faculties provided IT equipment and necessary hardware and soft ware facilities in 1998. With the higher   education councils restructuring attempt in education faculties in 1998, the teacher training curricular was revised and a new department in education faculties was created; and courses in ICT and its uses in teaching and learning were to be provided to improve the quality of teachers. The curriculum for each ICT in Teacher Education (ITE) program was reformed from theory-laden courses to more practice based courses (Alev, 2003). In the teacher training program, two courses are included, whose contents are: computer; instructional technologies and material development.

In view of comparative literature above, many countries around the world are developing policies and guiding teacher professional development so that they are trained and prepared to use ICT in pedagogy. UNESCO has played its role to produce guidance for less favored countries at the request of their governments. A World Bank report on ICT in education identifies ICT in teacher education and has funded it in some instances. A similar initiative remains unexploited in many third world countries. Usun (2009) concludes that despite clear need to prepare teachers for level expertise (that is, pedagogical application of new ICTs in teaching and learning), quite serious developments can be seen in the UK teacher training curriculum in ICT which seems to focus more on pedagogical skills of teachers, and not integration of ICT in pedagogy which needs to be improved. This is relevant to the current study where pre-service training in ICT integration is undertaken, but the application of the skill by practicing teachers is largely unknown.

Other countries like German, Japan and Turkish, ICT related courses still tend to be skills on how to operate hardware or software, while USA programs touch on pedagogy use in some phases of training. In addition, the foundation courses which provide an understanding and knowledge about philosophical and sociological underpinnings of ICT in education are missing in all the countries. The conclusions are valid for the current study in a developing country, still struggling to meet the demands of computing in schools in terms of teacher ICT literacy, quality hardware and software, which is directly linked to economic position of the country, that may not match the developed world, yet the needs are the same.

In the African continent, Unwin (2005) carried out a research study to explore the gulf between the rhetoric of those advocating the use of ICT in education in Africa and the reality of classroom practice. The objective of the study was to explore and report the outline and possible framework for the successful implementation of teacher training programs that make advantageous use of appropriate ICTs. The paper argued in its objectives that among six fundamental principles of good ICT practice must be addressed for such programs to be effective in Africa: a shift from an emphasis on "education for ICT" to use of "ICT for education"; an integration of ICT practice within the whole curriculum; a need for integration between pre-service and in-service teacher training.

The paper concluded with a framework for action to deliver the very real benefits of ICT for teacher training in Africa. The methodology was exploratory survey with generalizations drawn based on work in countries in Africa and elsewhere, notably China, Estonia, Ethiopia, Ghana, Kenya, Malawi, Mozambique, Senegal, South Africa and Tanzania (Unwin, 2005). Findings indicate that there are existing interventions of ICTs in teacher training in Africa; drawing primarily on ideas originating in Europe, Canada, the USA, Australia and New Zealand (Anderson & Elloumi, 2004; UNESCO, 2002b; Cabanatan, 2001; Jenkins, Shrestha, 2000; Lieberg & Stieng, 1998; and Somelch & Davis, 1997; and). There is necessity for teachers to first be trained in basic ICT skills (UNESCO, 2002b), after which there is need to address four competencies: Pedagogy, collaboration and networking, social issues and technical issues. The conclusion is relevant to the current study on training of teachers as focus is usually directed at basic skills of technology use; but inconsistent information is available on skills to use the technology in pedagogy.

Peter Kinyanjui (Kinyanjui, 2004) was NEPADs e-Africa program coordinator stressed that e-school initiative in Africa would ensure that majority of people on the continent have the skills required to function in the knowledge economy, and defined some NEPADs e-school objectives as follows: to minimize the effects of digital divide on young people and provide them with ICT skills necessary to function in the knowledge economy; to ensure that every African youth leaving school has the necessary ICT skills that will assist them to find jobs. Unim (2005) avers, the lack of mention amongst NEPAD's objectives of the use of ICT to enhance wider learning and educational experiences clearly illustrates that despite NEPADs increased rhetoric on teacher training, this initiative remains primarily about using education to enhance ICT skills, in the expectation or hope this will in itself be of benefit to African people.

The NEPAD initiative aimed to connect more than half a million primary and secondary schools in Africa to the Internet, but without comprehensive framework developed at national level to train teachers in the appropriate use of such technology in pedagogy, it was likely such initiatives would achieve little in the way of real educational change in the continent. Kinyanjui (2004) and others involved in the e-schools initiatives comment that teacher training is important; but, according to Unim (2005), until the core emphasis shifts away from focus on simply getting schools connected, to deeper understanding of how this can transform children's learning experiences, it will be doomed to fail. True to the prediction, NEPAD program never passed the pilot trial phase for full implementation in selected countries in the African continent.

In Kenya the final report of the task force on the re-alignment of the education sector to the constitution states that teachers are expected to put a deliberate effort to enhance their skills and capacity in the area of ICT integration (MOE, 2012), expected teachers to seek personal initiative to train in ICT, and went further to advice teachers to do all within their means to acquire the requisite ICT skills; to enhance their efficiency in service delivery and also empower them to play their rightful role in the envisaged knowledge economy (MOE, 2012). Practicing teachers responded by acquiring basic skills in computers but ignored the fact that ICT integration requires is more than just basic computer skills; it is about teaching with computers coupled with frequent exposure to, handling of and interaction with ICT equipment during pedagogy to enhance proficiency.

Unfortunately, teachers cannot apply what they have not been trained to do if the equipment is not accessible. Some schools in the republic have acquired computers through government grants or initiatives of their boards of management and parents association; and, introduced computer studies as a subject offered in their curriculum. Such schools have rooms set aside as computer laboratories. The current study targeted such schools to investigate if the computers are also utilized by teachers teaching other subjects like chemistry for integration during pedagogy.

The annual in-service training of all mathematics and science teachers by CEMASTEA from 2012 emphasize use of ICT in teaching and learning chemistry as a way of enhancing delivery and learning in the 21st century (CEMASTEA, 2018). The CEMASTEA team trained teachers to embrace ICT in teaching and learning. However, the missing link is the application of the skills by teachers in classroom instruction; and how the trained teachers would be evaluated without proper monitoring and evaluation strategies in place. This study assessed the extent to which practicing chemistry teachers trained by CEMASTEA or any other organization in Kenya have embraced the integration of ICT in teaching.

The specific topics of concern to the chemistry department during training included: the mole concept, organic chemistry and thermo chemistry. The CEMASTEA training activities involved the use of ICT tools such as radios, television (T.V.), camera, mobile phones and Internet; emphasizing the development of skills in the areas of Microsoft word, Microsoft excel, PowerPoint and Internet usage during integration in chemistry lessons (CEMASTEA, 2018). The training module also encouraged teachers to develop their own ICT materials using simple and available tools like mobile phones and video cameras. However, they ignored time allocation for such activities in the already overcrowded Kenyan education curriculum. This study's objectives of accessing chemistry teachers' capacity to integrate ICT link with closely the aims and objectives of CEMASTEA of training science teachers on ICT integration and assess the extent teachers are able to create and use digital content in chemistry instruction.

Mukuna (2013) researched on integration of ICT into teacher training and professional development in Kenya. The objective of the study was to respond to questions pertaining to challenges encumbering integration of ICT in the curriculum in Kenya; and attempted to respond to various questions on the issue of teacher training on ICT integration in pedagogy. Findings indicate incompetence of teachers and lack of training prior to introduction of new innovation is critical as the implementer and engine of the curriculum change. Success of an innovation depends on the characteristics of the innovation, context and content (Fullan, 1992, Otunga *et al.* 2011).

Mukuna (2013) recommends that teachers must understand how they are expected to fit in the changes requiring them to use ICT in their lessons; Hawkridge (2002), Hawkins (2002) also share the view that a good curriculum is invalid without quality teachers to implement it, and it should not be assumed that teachers know the change in curriculum and will find their ways around. Mukuna (2013) recommends that Kenya teachers should have Technological, Pedagogical content Knowledge (Mishra & Koehler, 2011); and, this should be done in their training and maintained through continuous professional development for teachers, administrators and technicians. Findings in the former study are relevant and binding to the current one; the recommendations are valid, and may be applied to make a breakthrough so much needed in the teacher training institutions in Kenya.

A more recent study by Mwangi and Khatete (2017) examined the teacher professional development needs for pedagogical ICT integration in Kenya focusing on lessons for transformation. A cross sectional and descriptive survey design was used as research data was collected through triangulation. Three key instruments for data collection used were questionnaires, interview guides and checklist. The study sample was thirty secondary schools from Nairobi and Kiambu County. Findings from the study revealed variance in the use of ICT by experienced teachers, especially between personal use and pedagogical use. Most teachers felt that the approaches used in professional development did not equip them adequately for independent ICT usage in pedagogy. The findings are binding to the current study; because when training focuses on teaching the technology instead of how to use the technology to teach, the end product is under utility by users coupled with fear of the technology and complacency of status quo. The finding is also valid and relevant to the present study as training develops skills, and practice infuses confidence in the users; it is important to establish how this applies to practicing teachers.

In an earlier study Isaboke (2014) researched on teacher preparedness in integrating ICT in instruction in lower primary schools in Nyamira County, Kenya. The main objective of the study was to find out the extent to which teachers training in ICT influence integration of ICT in teaching and learning. The study adopted a descriptive survey design. The target population was thirty four lower public primary schools. Purposive sampling technique was used to sample head teachers and lower primary school teachers. The sample size was forty one respondents comprising of ten head teachers and thirty one lower primary school teachers. Questionnaires and interview schedules were used as data collection instruments.

The study established that teachers' attitudes, training, teaching experience and teachers' level of self efficacy had a positive and significant effect on the integration of ICT in teaching and learning. The study concluded that teachers had low levels of ICT use for educational purposes. Nonetheless the study revealed that teachers had positive attitude towards use of ICT but were not ready to use them in pedagogy due to lack of appropriate skills and knowledge. Majority of teachers had been trained in the basic computer literacy at certificate level, but lacked competence on how to use ICT in pedagogy. The revelation from the former study is binding to the current study investigating skills in integrating ICT in pedagogy, for when application skills are under used or lacking in a teaching and learning environment the users hold back their experiences.

The study recommended need to train all teachers specifically on how to use ICT in instruction; asserting that this can be achieved by having teacher training curriculum with content on ICT integration in pedagogy. The researcher further asserts that availability of ICT infrastructure will not automatically lead to ICT integration in pedagogy. According to the researcher, there is need for the ministry of education to provide in-service training programs in ICT for teachers, especially those with more working experience; this will help them change their attitudes, equip them with ICT skills and increase their levels of self efficacy in ICT usage. This is in line with the current study as teachers need to enhance their ICT skills regularly and stay updated through continuous professional development.

In Kisumu County, Otieno (2018) investigated the influence of teacher preparedness on integration of digital technologies in early years of education in Kenya. The objective of the study was to establish the influence of teacher status of digital literacy on integration of digital technologies in early years of education. The study adopted concurrent triangulation design with mixed methods approach; and, targeted 345 teachers from Kisumu Central sub-county; adopting saturated sampling to sample 90 pre-school and 75 grade two teachers; stratified random sampling to sample 9 head teachers and 23 grade three teachers; and purposive sampling to sample 75 grade one teachers. Data collection instruments were structured questionnaires, interview schedules and focus group discussions. The findings showed a significant (n=202; r=.711; P<0.005) highly positive correlation between status of digital literacy and integration of digital technologies. The teachers' status of digital literacy was also found to be high. The study recommended that the ministry of education should enhance their supervision on the implementation of integration of digital technologies in classroom teaching. The findings are relevant, however the data collection instruments were flawed, as integration of digital technology can only be confirmed by observation; questionnaires and interviews alone tend to be prone to respondent bias and true picture may not be revealed.

The ICT literacy of teachers as implementers of curriculum is important for the process of integration to be successful. There is inconsistent information on whether emphasis is on teachers training on ICT use in pedagogy or whether emphasis is on use of the technology in pre-service training and in-service capacity building for practicing teachers in Kisumu County. Why performance in chemistry has remained low consistently over the years even in schools with computers and yet ICT integration is supposed to simplify difficult and abstract chemistry content as a subject needs to be established. These are some of the pertinent questions that need to be addressed by the study.

2.4 Extent of Integration of ICT Resources in Public Secondary schools

Globally, resource constrained contexts are considerable, and, therefore affect the extent of integration of ICT in teaching and learning thus: training of teachers and administrators, connectivity, technological support and software amongst others (Barak, 2007). Schools in some countries around the world in order to ease burden of supply of ICT equipment, have began allowing students to bring their own mobile technology such as laptops, tablets or smart phones into class, rather than providing such tools to all students; an initiative called "Bring your own device"(Denisia & Suresh, 2013). Such initiative however, may result in inequality as some parents may not afford the same for their children; under such circumstances schools are tasked to ensure equality.

Kisirkoi (2015), Bonfilius (2019); Sangra and Gonzales (2016) (MOE, 2012); Alvaro (2011); and Paspia (1994) define extent of integration of ICT resources as the degree of using any technological tools such as Internet, e-learning technologies and CDROMs to assist teaching and learning; extent of use of ICT in education has been reported to result in many learning benefits though it is quite demanding; asserting that, on one hand, enough preparation has to be put in place for it to succeed. According to the researchers, ICT integration in education requires keen planning, effective teacher preparation and sustained regular teacher professional support and visionary leadership that recognize the need to prepare the learners to live and work in the technological world of 21st century. As Sangra and Gonzales (2016) put it, extent of ICT integration requires teachers' attitude that is adaptive to change and appreciation of the fact that in many ways modern technology enriches pedagogies. The researchers recommend that in order to appreciate integration of ICT for improved educational quality, both technology and pedagogy must be addressed. ICT has the potential to "bridge the knowledge gap" (Lu, Hou & Huang, 2013) in terms of improving quality in education, increasing quantity and quality of educational opportunities, making knowledge building possible through borderless and bondless accessibility to resources and people, and reaching populations in remote areas to satisfy their basic rights to education.   This motivated the current researcher to seek to establish preparedness of schools on ICT integration to bridge the gap in Kenya.

Spiro (1996); Kolb (1991); Flavel (1979); and Schank (1991) view ICT as a source of motivation (stimulant) that is a necessary component in learning, because it causes the learners' sensory apparatus to be activated (stimulant). Relevance, curiosity, fun, accomplishment, achievement, external rewards and other motivators (stimulants) facilitate ease of learning (response) in the stimulus and response theory. This is addressed in the current study where ICT integrated teaching environment is the stimulant and learner performance is the response.

Mishra and Koehler (2011) developed a theory to describe integration of ICT in teaching and learning known as Technological Pedagogical and Content Knowledge (TPACK) framework. The researchers posit a teacher depends on three domains of knowledge for effective integration of ICT into teaching and learning: the domains are Content Knowledge (CK), Pedagogical Knowledge (PK) and Technological Knowledge (TK); and defined them thus: Content Knowledge is the knowledge about the actual subject matter   that is to be taught and learnt. Professional qualification for a teacher is subject content knowledge. Pedagogical Knowledge is the deep knowledge about the process or methods of teaching and learning (for example objectives and aims, classroom management, lesson planning and student evaluation). The researchers argue that a teacher with a deep PK is likely to integrate technology in his or her teaching, considering how learners can best learn in a given classroom context and nature of learners. Technological Knowledge (TK) is defined as knowledge about standard technologies. They assert that a teacher with TK has good knowledge of operating systems and computer hardware, the ability to use standard sets of software tools (for example word processors, spread sheets, browsers and e-mail) and to install and remove peripheral devises, install and remove programs, create and archive documents among others.

Therefore, according to Mishra and Koehler (2011) TPACK is the intersection of all the three bodies of knowledge (CK, PK and TK). The researchers argue that the development of TPACK by teachers is central for effective teaching with technology. In view of the issues raised above, any effective implementation of technology in the classroom therefore requires acknowledgement of the dynamic transactional relationship among content, pedagogy and the incoming technology, all within the unique context of different schools, classrooms and cultures. From the theory therefore, factors such as the individual subject teacher, specific grade level, the class demographics and more mean that every situation will demand a slightly different approach to the integration of technology in education, hence specific lesson plan.

The extent of ICT integration in teaching and learning chemistry in Europe was presented in a report by Morante, Lopez, Fernandez and Miranda (2014). The report indicated that since 1998 the European governments have invested some £ 1.8 billion in the national grid for learning and teacher training with the aim of helping teachers use ICT integration to raise standards, performance and transform teaching and learning. The researchers report (on the study carried out between the years 2008 to 2009 in five European countries: United Kingdom, Spain, Austria, Hungary and Denmark) that 99% of secondary schools have network, are connected to Internet, have broad band connections and interactive white boards. The average number of desktops available is about 246 per school; while the student to computer ratio was averagely 3.6 pupils. According to the report, research indicates that almost 88% of secondary schools are connected to the Internet and classrooms have computers and access to Internet is high at 82%. This contrasts with the current study conducted in a developing country where computers are a handful when available, and are only available at the ICT center; these cannot suffice to serve integration of different subjects by different classes unless timetabled effectively; at the same time the learner computer ratio is averagely 150 students (MOE, 2012).

Morante *et al.* (2014) report that despite the wide diffusion of ICT and the high connection rates, European schools tend to use the technology primarily for presentational and display purposes in teaching subjects like chemistry, rather than innovative teaching and learning programs. This point is underplayed by Sanchez and Aleman (2011) who note that the use of ICT beyond basic level is fairly limited in European schools; the full potential of ICT as educational tool will remain unrealized if they remain merely for presentation or demonstration, yet this is what is evident in the former study.

The report by Morante *et al.* (2014) state that unless teachers have high level of skills they do not always have time to explain and implement ICT in any depth in their teaching. Nonetheless, the report indicates that 81% of secondary teachers feel confident about using ICT. Though there is an increase of teachers uploading and storing information for use in lesson planning; majority still use paper-based resources with about 50% who use self-created digital resources to plan their lessons. The finding in the report is relevant to the current study as it shows that even in Europe the extent to integrate ICT in teaching and learning in secondary schools by teachers is at low levels and is relegated more to planning, presentations and display purposes. All these studies indicate basic use of ICT in European schools despite the availability of infrastructure and equipment, there is need for a comparative study in developing countries.

In Asia Shahid (2015) carried out a study with an intention to understand the extent to which Asian science teachers are equipped with ICT skills and their self confidence in the use of ICT. The objective of the study was to highlight the concern for the level of ICT application skills and confidence of science teachers in Asia. The study randomly selected a sample of one hundred and ten trained science teachers from different countries of Asia working in Malaysia, Maldives, Sri Lanka, India and Singapore. The study design was survey, using structured questionnaires developed to determine the application skills along with focus on technical skills and attitudes towards ICT, using a model with three main areas: Teachers' basic ICT skills and self confidence, teachers' application skills and self confidence and teachers' attitudes towards ICT.

From the findings the study identified that application skills of science teachers in Asia was significantly low when compared with training they had acquired. This was also identified in an earlier study by Higgins (2010) that ICT application have got some problems one of which is negligible use of it; the researcher stated further that unless there is effective use, having more computers does not make any difference. Shahid (2015) study also identified that though the respondents were from different cultures, as teachers, their application level had remained almost the same without any difference on gender. The researcher concluded thus, because Asian teachers have positive attitude towards ICT integration in teaching and learning, and because the purpose of promoting ICT in schools is to increase effectiveness of teaching and to acquire progress for the learning approach of students; he recommended that it is necessary to provide further ICT training for Asian teachers and to include compulsory module of ICT application in the training;

recommending further facilities and opportunities for the teachers to participate in processes where ICT is the core area.

The finding is relevant to the current study which sought the extent of ICT integration. The former study however did not establish the root cause of the negligible use of ICT by the Asian teachers; the conclusion and observations are valid. Further training and compulsory modules recommended in the former study can be comparable to CEMASTEA in-service courses (CEMASTEA, 2012), which include ICT integration offered to science and mathematics teachers in Kenya. These courses can be regarded as professional development for mathematics and science teachers.

In South Africa, Pandayechee (2017) carried out a survey on the level of ICT integration in South African schools. According to the researcher, extent of ICT integration in education in South Africa had been severely limited by operational, strategic and pedagogical challenges. The objective of the study was to determine the level that use of ICTs (digital media, e-learning tools, online services and digital devices) have been integrated into the pedagogical and content knowledge as advocated by Mishra & Koehler (2011) theory in South African schools. A non-experimental exploratory survey methodology was designed for the study. Purposive sampling of teachers was carried out across all disciplines of which 28% were mathematics and science teachers from 34 secondary schools in Tshwane South. The purposive sampling criteria considered a confluence of relatively high access to the Internet and top performing secondary schools, as this was expected to generate best case scenario of ICT integration in education in South Africa.

According to the researchers, criteria for purposive sampling were based on: Wi-Fi sites around the school (Jack, 2016); the school was considered most advanced with respect to installations and uptake of ICT (Gilbert, 2015). The former researcher argued that access to the Internet connectivity would encourage higher uptake of ICTs in the region. The second criterion was selecting high achieving schools based on 2015 database. Tshwane South contained the largest proportion of schools that performed above the provincial average of 84.2% in the national senior certificate. The data collection instruments for both qualitative and quantitative data was a questionnaire with two structured questions with eighty items each, adopted from Zawescki-Richter *et al.* (2015). The technical knowledge was supported by open-ended questions from Graham *et al.* (2009). According to the researcher, the open-ended questions allowed the participants to freely voice their concerns and make suggestions.

From the findings, the study established that the extent of uptake of technology is low in South Africa. On average the frequency of usage per tool was: 41% on contextual tools, 29% on sharing information and ideas tools, 18% on reflective dialogue tools. It was established that teachers were uncertain with respect to enforcement of ICT integration in education while being encumbered by poor infrastructure and lack of skills. Pandayechee (2017) in his conclusion described it as a misconception that merely providing technology can transform education. He cited challenges that lie not only on how to use the technology, but also on how to integrate digital technologies effectively into the curriculum. The findings and conclusion are relevant to the current study. The localization of the study, coupled with the sampling technique gives a narrow scope and limited external validity of the research. In the current study triangulation was coupled with observation to strengthen the findings.

In Eastern Africa, Abdelrahman, Howre and Osman (2013) investigated the level of integrating ICT in teaching and learning mathematics and sciences in Sudanese secondary schools in Khartoum in relation to a number of selected countries (Chile, Slovenia and South Africa). Objectives of the study was to ascertain the extent which ICT is integrated with teaching programs of science (chemistry and physics) and mathematics teachers at secondary schools, and to determine obstacles that prevent integration of ICT in teaching process at Sudanese secondary schools. The study employed a survey methodology, such as SITES modules for Secondary ICT in Education Study (Pelgrum & Anderson, 2001). Stratified sampling of fifty secondary schools relatively advanced in using ICT for educational purposes was sampled. The study used a questionnaire to collect data from science and mathematics teachers; data was collected by personal administration approach.

Findings by Abdelrahman *et al.* (2013) indicated that: majority of teachers in the sample schools did not integrate ICT or use it to assess learners in science and mathematics classrooms; and none of the schools used computers in teaching and learning scientific activities for their students after scheduled school hours and majority of principals did not support their teachers in integrating ICT in teaching and learning process. According to the findings, these observations were attributed to: lack of training courses for teachers to use ICT; limited time for planning; examination pressure; fear of not being able to complete syllabus; inadequate infrastructure; lack of Internet; absence of any kind of ICT management system in most schools and negative attitude of some teachers. Finding is relevant to the current study as ICT type of training of pre-service and in-service teachers is important in determining their extent of integration of ICT in pedagogy, yet it is largely unknown.

The research concluded that Sudan is behind many countries internationally in integrating ICT in education and does not have the necessary infrastructure for the process; despite that most teachers had positive attitude regarding ICT integration in teaching and learning. Abdelrahman *et al.* (2013) aver although the ICT implementation policy for Sudan was launched in 2002, and most schools have computers and Internet connectivity, most principals of high schools, teachers and students do not really know what to do with the computers installed in their labs; showing lack of careful planning. And, the researchers aver further, the implementation was top-down initiatives which did not take into account the involvement of local policies; recommending support and guidance. The conclusion is relevant to the current study in the context where computers available at selected institutions are dedicated to computer studies as a subject; nonetheless, integration of the technology to teach other school subjects is largely unknown in Kisumu County.

The researchers quote Pelgrum and Law (2009) who did a similar research and concluded that top ministry leaders in Sudan down to teachers in their classrooms, all face decisions about whether and how to integrate ICT in teaching and learning a new domain in Sudanese schools. The research finding was relevant and conclusions binding to the current study. Nonetheless, the sample size was too small for generalization to the wider population in Sudan and rest of Africa. The method of data collection was faulty as the questionnaires alone are prone to subjectivity and may not give the correct picture objectively. In the current study triangulation and a follow-up with observation are used to fill the gaps that serve as confirmatory to the teachers responses which may have been subjective.

In Kenya, Kisirkoi (2015) did a case study to investigate level of integration of ICT in education in Kenyan secondary schools. The researcher was inspired by one secondary school; despite reports that there was very little integration of ICT in curriculum delivery in many secondary schools in Kenya. According to the researcher it was reported that the school was using ICT in instruction, and was practicing learner centered instructional approaches and there was improved learning outcomes and performance in national examinations, the mean score in KCSE improved from 6.2 to 8.4 between 2007 and 2013. The objectives of the study were to: investigate teacher computer literacy levels, motivation for integration, perceived reasons for the intervention and the impact on teaching and learning.

The study design was case study with a population of five hundred and thirty five students and twenty eight teachers (Kisirkoi, 2015). Simple random sampling technique was used to sample thirty students and eighteen teachers. It was established that students and teachers in the study school were computer literate and were able to manage computer applications in teaching and learning. The learning process was found to be practical with learner interactions and activity based learning. The motivation was desire to teach better coupled with visionary, supportive school leadership. The school was using ICT as a teaching learning tool and there was improvement of learning environment and outcomes. The study established that the school had an ICT policy thus: any new teacher who joined the school must attend evening computer literacy classes for two months to acquire basic computer literacy skills; after which the teacher trains his or her students on computer use, the students in turn teach one another after classes how to work with computers in the computer lab with the teacher on duty supervising. Teachers post students assignments in the school website. Teachers learn from one another how to gather materials from the web and customize it to use in their lessons. They cast concepts on the screens and students learn better with presentations and pictures; making work easier and enjoyable. This is the scenario that can be referred to as ICT enhanced school environment.

The case study by Kisirkoi (2015) concluded that ICT integration in instruction was of benefit both teachers and learners. The researchers recommended that other schools should emulate this school. The findings in the study are binding to the current study on level of ICT integration in public schools; and conclusion is valid. However, localization of the case study to only one school limits its generalization to other secondary schools in Kenya.

Karenji (2016) investigated integration of ICT in teaching English in secondary schools in Nyakach sub-county, Kisumu County. The problem of the study was that only a few schools have embraced the use of ICT in teaching and learning despite its enormous benefits in everyday life in and out of school. The study sought to investigate the extent to which teachers were using ICT in teaching and learning of English in secondary schools in the sub-county. The study adopted descriptive survey design; the sampling technique was a combination of stratified, purposive and random sampling procedures. The sample consisted of 22 secondary schools with ICT facilities, 7 school principals, 16 English teachers and 540 Form three students. Questionnaires and observation schedules were used to get information from respondents. Findings showed that the use of ICT in teaching and learning of English was still in the formative stages and faced various challenges; the available ICT resources were only occasionally used. The study recommended that ICT be fully integrated in the education system, and intensive resource mobilization be put in place to enable schools acquire ICT resources. The findings are relevant and binding to the current research. The recommendations are valid. The study researched on English is likely to limit its generalizable to other subjects. The current study in chemistry can be used to bridge the gap.

According to National Research Report-Kenya, one of the impacts of ICT on teaching and learning is clarified concepts in the sciences and mathematics and more interesting presentations to the learners to impact on improved performance (ROK, 2012). Because PowerPoint presentations jog learners' minds; the paper recommends major policy dialogue requiring attention such as the modernization of classrooms to accommodate ICT. Contrary to the developed world, classrooms in public schools in developing world, Kenya included are not yet automated to accommodate the technology. However, for the schools targeted in this study, ICT is available in the computer labs or ICT centers; it was the objective of the study to ascertain the extent the equipment is available, accessible and can be integrated in teaching and learning.

Research by CEMASTEA (2012) recommends use of ICT by teachers to attract students' interests and attention, ICT integration makes learning meaningful (CEMASTEA, 2018) in chemistry ICT has been proven to work best in peer coaching and teaching by teachers providing problems that demand analysis, evaluation or synthesis and such problems should have multiple solutions. The researchers' advice teachers to encourage students to participate in finding the solutions, this enhances understanding of the content and higher order thinking skills. The recommendations are in concordance with the current study on extent of integration of ICT, addressing increased student participation and motivation, which leads to improved performance and willingness to learn chemistry; factors that have changed traditional belief that science subjects are difficult. According to the former research, ICT attracts learners' interest and attention and enhances understanding. But without proper guiding policies on ICT integration in Kenya nationally or at school levels, information on the extent of ICT integration by teachers' to teach school subjects, specifically chemistry in Kisumu County remains largely unknown. More important is to determine the extent that the equipment is accessible, adequate, utilized and influences performance of school subjects such as chemistry.

2.5 Stakeholders Perceptions of ICT Integration on Learner Performance

Globally, there is widespread perception that ICTs can and will empower teachers and learners, transforming teaching and learning process from being highly teacher-dominated to student-centered; and that this transformation will result in increased learning gains for students, creating and allowing for opportunities for learners to develop their creativity, problem solving abilities, informational reasoning skills, and other high order thinking skills (Thayer, 2020). However, according to the report, there are currently very limited unequivocally compelling data to support this.

Earle (2002) holds the view that specific uses of ICT can have positive effects on student achievement when used appropriately to complement teachers' existing pedagogical philosophies. Educational systems are also adopting the new technologies to integrate ICT in teaching and learning process, to prepare students with knowledge and skills they need in their subject matter (Hussain *et al.*, 2017); in this way the teaching profession is evolving from teacher-centered to student-centered learning. This motivated the current researcher to investigate the perception of Kenyan stakeholders on integration of technology in teaching and learning chemistry in Kenyan schools to establish their views.

Cadellini (2012) investigated the aspect common in every culture as the decreasing number of students studying chemistry and the barriers that prevent students from studying chemistry. The researcher states that for many students, chemistry is seen as a difficult, complex and abstract subject that requires special intellectual talents and too much effort to be understood. Nonetheless, the researcher points out that more than other sciences, understanding chemistry relies on making sense of the invisible and untouchable; noting that the source of students' difficulties can have at least three origins:

1. The nature of science itself makes chemistry inaccessible.

2. The methods by which teachers have traditionally taught chemistry raise the problem

3. The methods by which students learn are in conflict either with the nature of science and teaching methods or with both (Cadellini, 2012).

In Kenya, chemistry is a compulsory science at secondary school level, learners are obliged to study it at this level, and mass failures are experienced all over the country; there is void information on any attempt to improve performance of learners in the subject.

Dori *et al.* (2013) researched and presented a perception of using and incorporating ICT into teaching and learning chemistry. Investigation on students ICT skills in chemistry in particular, and sciences in general and established that ICT based learning environments play a significant role in education. Another research by Barnea *et al.* (2010) has exemplified visualizations in science laboratories such as molecular modeling, data collection and presentations; while Bell *et al.* (2010) in their study focused on ICT use via World Wide Web (WWW) and virtual reality, as well as the roll of ICT for developing higher order thinking skills such as inquiry, graphing and modeling. Other examples given included different assignments for teaching chemistry using ICT. Internet connectivity is a challenge in most Kenyan schools for access to the Web and virtual reality.

Muhammad *et al.* (2019) investigated the impact of ICT on students' academic performance in chemistry by applying association rule, mining and structured equation modeling. The study established that ICT plays a significant role in students' academic performance. From statistical and mining perspective, overall results of descriptive statistics, reliability analysis, confirmatory factor analysis, OLS regressions, structured equation modeling and data mining algorithms such as association rule, mining and éclat have been employed to evaluate the comparative importance of the factors in identifying the academic performance of students. Overall results indicate that there is a significant relationship between ICT use and students academic performance. The proposition suggests a significant relationship between ICT integration and learner performance, however they ignore the distinct effects that raise questions about effectiveness of educational policies that guide ICT use in education.

Based on statistical analysis, it came to light that ICT positively affects students' academic achievement and retention; and, ICT was found more compelling, effective and valuable in teaching chemistry when contrasted with conventional techniques of teaching. Hussain *et al.* (2017) investigated the effects of ICT on students' academic achievement and retention in chemistry at secondary level. Fifty students of ninth grade were selected randomly from Kohsar public school and college, Latamber Karak. The students were grouped into equivalent groups based on pre-test scores. Pre-test post-test equivalent groups design was used. Findings based on statistical analysis indicated that ICT positively affects students' academic achievement and retention. ICT was also found to be more compelling, affective, rewarding and valuable in teaching of chemistry when contrasted with conventional techniques of teaching. The study recommended that ICT should be used in teaching chemistry for enhancing students' academic achievement at secondary school level. This finding motivated the current researcher to investigate the perception of stake holders in Kisumu County on their views of ICT integration.

Davis and Tearl (1999) hold the view that ICT has the strength to speed up, improve and extend aptitude reforms as it has the capacity to boost teaching by inspiring and engaging learners. Ashley (2016) reiterates that technology helps educators in preparing students for real world setting and stresses that as a country turns out to be more technologically dependent, it becomes significantly essential for students to figure out how to be well informed about ICT. Badeleh and Sheela (2011) inferred that generally to study chemistry, component based achievement, retention of learning and comprehension are required, and, ICT was more successful than the laboratory training model of teaching. The propositions are relevant and binding to the current study as the subject chemistry has experienced mass failures in Kenya that have not been attended to. There is insufficient information to explain reasons behind the poor performance in chemistry in Kenyan schools. It is the policy of the Ministry of Education (MOE, 2006) in Kenya that teachers should integrate ICT in teaching and learning using even the basic infrastructure including Internet and cell phone. The current research sought to investigate the perception of stake holders on ICT integration in teaching chemistry.

Avinash and Shailja (2013) established that ICT program is more compelling and effective than the conventional teaching approach in terms of students' achievement scores in chemistry. This view is shared by Oginni and Popoola (2013) who established that by using ICT students' retention scores were better when contrasted with those who were instructed via conventional methods, affirming the results of former researchers that students learn and retain better when they are taught through ICT. This finding is relevant and links to the current research on ICT integration in teaching and performance in chemistry.

Lindfors (2007) investigated the perceptions of a group of European chemistry teachers on the use of ICT in teaching; under the research question: what are the perceptions of European teachers on the use of ICT in education? The objective of the study was to highlight the challenges and goals of the use of ICT in teaching for the European information society. The research data was collected in the frame of Future In-service Teachers Training (FISTE) project (FISTE, 2004-2007). The analysis was based on European teachers' discussions. Teachers from different parts of Europe in five groups discussed the pedagogical use of ICT and its impact on performance. The findings revealed that four main categories of teachers' views were formed: the value of using ICT in teaching; the ICT competencies of teachers; the pedagogical challenges of using ICT in teaching; and the future of ICT in pedagogical use. According to the findings, the value of using ICT in teaching formed the largest category of teachers' views; here, the view was that teaching was the most important thing and ICT as a tool or method was on the second place. The teachers recommended a balance between ICT and traditional teaching. Lindfors (2007) made conclusions as an answer to the research question of the article as follows: Teachers share the idea that technology itself cannot create new types of teaching. According to the teachers, technology has to be used and developed on the basis of pedagogical ideas; new ways of communication can create opportunities for solving global problems. From the findings in the former study, even European teachers have perceptions that use of ICT has to be combined with traditional classroom based teaching from pedagogical point of view.

According to the teachers' perceptions, the impact of ICT in education greatly depends on how it is used. However, the teachers seemed to need support in developing the use of ICT more in the collaborative direction. The research findings are relevant and binding to the current study and the teaching fraternity in general, because universally teachers face similar challenges and share similar experiences. Nonetheless the infrastructure available in European schools may not apply in the same context universally, and this limits generalization.

In Namibia, Jatileni and Cloneria (2018) researched on teacher perception on the use of ICT in teaching and learning: a case of Namibian primary education. The objective of the study was to investigate the perceptions of stakeholders such as teachers hold towards the use of ICT in teaching and learning as key determining factor to the success or failure of the use of ICT in education; the study further aimed to explore the extent to which Namibian primary school teachers use ICT in their classrooms and the criterion they based on the use of ICT for teaching.

Self administered questionnaires with open-ended and closed-ended questions were anonymously used to collect data from primary schools across Omusati region in the northern part of Namibia. A total of ninety teachers participated in the content of the qualitative data that was descriptively and inferentially analyzed using SPSS software. Qualitative data was discussed along with quantitative data analytics. Findings indicate that stake holders agreed that ICT usage in schools will improve teaching and learning; and indicated that there was moderate use of ICT in classrooms (Jatileni & Cloneria, 2018).

The results showed that teachers decide to use ICT in teaching and learning based on: lesson objectives, activities, subject policy, curriculum learners diverse learning needs, accessibility and availability of the ICT devices. The results indicate views of the teachers, the methodology and mode of data collection did not give chance for the researcher to confirm these views which would have been done by classroom observation to confirm that the teachers practice what they stated. The instruments gave room for respondent bias without observation, which would have strengthened the outcome of the study.

In Rwanda Munyengabe, Yiyi and Hitimana (2017) investigated primary teachers' perception on ICT integration for enhancing teaching and learning through the implementation of One Laptop per Child (OLPC) program in primary schools in Rwanda as the main target of the study. The study employed qualitative approach where thirty primary school teachers participated in the study through group discussions designated for the research questions, which were related to the benefits of ICT in education: requirements to integrate ICT into teaching and learning practices; challenges hindering the implementation of OLPC program; and the contribution of different stakeholders for the implementation of OLPC program in primary schools of Rwanda. The main research question was: what would be the teachers' perceptions on benefits from the integration of ICT and the implementation of OLPC into teaching and learning process in primary schools of Rwanda; what would be the stakeholders' contributions in integration of ICT through implementation of OLPC program in primary schools in Rwanda (Munyengabe *et al.*, 2017).

Data was collected in two phases for each category of participants. In the first phase the researcher and research assistants attended different trainings where it was possible to meet different teachers familiar with ICT use to collect data from group discussions. The second phase was to conduct in-depth interviews with all volunteer teachers who participated in the study. The second phase data was collected by interviews after group discussions. In the group discussions, different groups of ten teachers each were grouped and given materials to be used while making a draft of what they were identifying.

From the results, data collected from the groups allowed the researchers to identify two categories of teachers: the first group are those teachers who are usually using ICT and are aware of how to use ICT tools; the second group are those teachers who are interested to use ICT because they have heard or acquired the importance of using ICT in teaching and learning process. The teachers' perceptions on ICT show their interest to integrate ICT into teaching and learning process. In both categories teachers believe to benefit from ICT by sharing experiences, advice and expertise. According to the findings, teachers are aware that ICT will help learners do their self coaching and that teachers can use the technology to illustrate and demonstrate new content. Findings show that teachers have positive feelings towards ICT integration into teaching and learning process in primary schools of Rwanda. Nonetheless, the findings indicate that Rwandese teachers are facing challenges relating to lack of adequate skills required to integrate ICT into teaching and learning process, lack of motivation due to financial constraints; and teachers asked the government for support in form of training and infrastructure. Findings also indicate that teachers recognize the contribution of stake holders in the OLPC program. The researchers concluded that the study enlightened teachers' perceptions for integrating ICT into teaching and learning process, by incorporating OLPC program in primary schools in Rwanda. Munyengabe, *et al.* (2017) recommended that all stake holders should participate into the OLPC program by contributing in solving all challenges encountered in everyday lives within schools. The findings are relevant in revealing the plight of teachers, who need motivation and support, especially where learners are the benefactors and teachers the implementers of the program. The success of the program depends on teachers perceptions.

In Kenya, Mwendwa (2017) investigated the perception of teachers and principals on ICT integration in the primary schools in Kitui County, Kenya. The main research question of the study was: what are the perceptions of teachers and principals on ICT integration in public primary school curriculum in Kitui County, Kenya? The study adopted mixed method research approach. A sample of three hundred and eighty eight principals and seven hundred and seventy six teachers was used for the study. The sample was drawn from three hundred and eighty eight public primary schools in the county, selected by stratified and simple random sampling method. Purposive sampling method was used to sample principals while the teachers were selected through simple random sampling method. Data was collected using questionnaires for teachers and interview guides for the principals.

The results indicated that most teachers had positive perception towards the use of ICT hardware and software tools in their instruction process. Through interview schedules, the principals revealed that the utility of ICT in the instructional process can greatly improve time management in schools, a pointer to positive attitude towards integration of ICT in teaching and learning in schools. The findings revealed that teachers and principals had positive perceptions on ICT integration in the curriculum, if only the required resources and facilities for ICT integration in the curriculum were present in their schools (Mwendwa, 2017). The study recommended that the government should continue with its efforts on ICT integration in education given the positive perception of teachers on the benefit of ICT in instruction process. The findings are relevant and binding to the current study, the conclusions and recommendations are valid; however the former study sought the perceptions of primary school teachers and head teachers that may limit its generalization to secondary school teachers and principals.

A parallel study in Kenya by Wasike (2018) investigated the perceptions of teachers, learners and school principals on the integration of ICT in teaching and learning secondary school agriculture in Bungoma County, Kenya. The purpose of the study was to establish the teachers' perceptions on the use of ICT in teaching and learning of agriculture in Bungoma County. The objective of the study was to ascertain the teachers' perceived usefulness of ICT, ease of use and adoption of ICT and their preparedness to use ICT as a pedagogical tool in secondary schools in Bungoma County; the study also gathered perceptions of learners and school principals on the use of ICT in teaching and learning of agriculture.

A descriptive survey research design was employed in Wasike (2018) study. The target population of the study consisted of all Form three agriculture students, four hundred and ninety eight agriculture teachers and two hundred and fifty two principles of secondary schools in Bungoma County. Purposive sampling technique was used to select sixty five principals, one hundred and twenty secondary school agriculture teachers and seven hundred and eighty Form three agriculture students as respondents. A questionnaire was used to collect information from agriculture teachers and school principals. An observation checklist was use to ascertain state of ICT in the schools. A discussion guide was used for the focused group discussion with students of agriculture. Quantitative data was coded and analyzed using SPSS. Qualitative data was analyzed using document report analysis. Findings indicated that perception of agriculture teachers, principals and students towards the use of ICT in teaching agriculture was positive; a few teachers used ICT in teaching agriculture compared to other subjects. The study recommended formulation of policies that promote integration of ICT in teaching and learning; and recommended teachers to use ICT in teaching and learning (Wasike, 2018). The finding is binding to the current study and reveals a gap on integration of teaching different school subjects in Kenyan schools. Nonetheless positive perception may not necessarily translate into application of the technology in practice; without physical lesson observation.

Research shows that there is a significant relationship between ICT integration in education and academic performance of learners. Communication about an innovation can create awareness among stakeholders and consequently create positive perceptions of the technology. This is because in Kenya, the allocation of time for teaching chemistry is strictly four or five lessons per week for lower and upper secondary respectively. With demand for performance, overcrowded syllabus and limited staff; teachers' integration of ICT in pedagogy depends on several factors including their work load, school enrollment, availability and access to the ICT center and their training to support its use. There is insufficient information on perspective of teachers in Kisumu County on the influence of ICT integration on chemistry performance. The current study sought to establish stakeholders' perspectives of ICT integration on learner performance.

2.6 Summary of Literature Review

The first part of the literature presented definitions and theories of Information Communication Technology (ICT) as a tool and as a resource for teaching and learning globally and then locally and its effects on learner performance; it also reviewed the genesis of ICT integration and national ICT policy in Kenya. The section that presented literature on level of integration of ICT resources examined rationale and theories for integrating ICT in education and stressed on need to train teachers not only on technology use, but to use technology to teach school subjects, hence need to have computers in the classrooms or in science laboratories; of which in parts of Africa and in Kenya the implementation status remain unaccomplished. The section that presented literature on chemistry teachers' capacity to integrate ICT indicates that studies agree that for teachers to successfully integrate ICT in pedagogy, they need professional development. On stake holders perspectives globally, teachers aver ICT use has to be combined with traditional classroom based teaching and its impact in education depends on how it is used, but agree that ICT integration is a booster for academic performance.

On the whole, the literature revealed some gaps as far as ICT integration in teaching and learning chemistry among public secondary schools is concerned:

a) Efforts of different governments and administrations globally focus on providing schools with good educational equipment; in Kenya however analysis of the available ICT equipment in computer labs for educational integration of ICT in the classroom to teach specific subjects like chemistry is lacking.

b) Studies indicate that quality professional development and ICT training of teachers is the root to any successful technology program. Despite numerous plans globally to use technology in schools, teachers have received little training through teacher education programs. There is need to establish whether the emphasis often on teaching about technology instead of teaching with technology results in adequate preparation to use the technology by practicing teachers.

c) Without proper guiding policies on ICT integration in Kenya nationally or at school levels, there is inconsistent information on the level ICT integration by teachers to teach school subjects, specifically chemistry.

d) Research show that there is a significant relationship between ICT integration and academic performance of learners in chemistry. Communication about an innovation can create awareness among stake holders and consequently create positive perception of the technology. How teachers value integrating ICT in teaching and their perception of the technology on their learners performance need to be established.

CHAPTER THREE
RESEARCH METHODOLOGY

3.1 Overview

This chapter gives a description of the research design, study area, study population, sampling techniques and sample size, instrumentation, data collection procedures, piloting and methods of data analysis.

3.2 Research Design

This study used a combination of descriptive survey and sequential exploratory designs. These were used to allow the researcher gather information about the current state of ICT integration in schools, and then make follow up and draw valid general conclusions from the facts discovered. According to Gall, Borg and Gall (2007), Orodho, (2005), descriptive survey research is the description of the state of affairs as it exists at present. The design provided an opportunity for the researcher to probe deep and obtain precise and concise information about the study population; and, gather information about the present and existing condition of the phenomena under study.

Sequential exploratory mixed methods design according to Ivankova, (2011), Ivankova, Vicky and Clerk (2005), implies collecting and analyzing quantitative and qualitative data in two consecutive phases within one study. This design is characterized by an initial quantitative phase of data collection and analysis, followed by a phase of qualitative data collection and analysis, with a final phase of linking of data from the two separate strands of data. This design was used in this study because it enabled the researcher to validate quantitative results with qualitative findings by using a lesson observation checklist to attend ICT integrated lessons of teachers who had indicated tendency to integrate ICT in their chemistry pedagogy. During which time the researcher visited selected study schools and did lesson observation in ICT integrated chemistry lesson. The results were related to the outcomes from the first quantitative data.

3.3 Study Location

The study was carried out in Kisumu County which is one of the devolved units of Kenya. It is located in Nyanza region and covers 2,119.10 square kilometers. It is 1,131 meters above sea level; the coordinates are $0^06'$ South $34^045'$ east. Kisumu County is bordered by Vihiga County on the Western side, Kericho County on the Eastern side, Kakamega County on the Northern and Homabay County on the Southern; while, Siaya County is located on the Southwestern side. Its headquarters is Kisumu City and consists of seven constituencies namely: Kisumu East, Kisumu West, Kisumu Central, Seme, Nyando, Muhoroni and Nyakach. The County has five local authorities: Municipal council of Kisumu, County council of Nyando, County council of Muhoroni and Town council of Ahero. Kisumu city is the administrative capital. The county is known for its association with Lake Victoria, the largest lake in Africa (Appendix Q: Map of Kisumu County).

The reason the researcher chose Kisumu County as the study location was within the concepts of the intention to improve performance in chemistry at KCSE, being a county where the average mean for KCSE chemistry from 2013-2016 has been 4.45 (grade D) out of 12.00, one of the lowest in the region. Yet, Kisumu County is a potential hub for chemical industry in the region, with industries like Agrochemical, Muhoroni sugar factory and Kenya Sugar Research. There is potential for a discipline like chemistry, a career subject yet very abstract, featuring in the national ranks at KCSE by diversifying pedagogy to include ICT integration in teaching and learning to make the subject less abstract and more practical and learner oriented.

3.4. Study Population

For this study the schools that were picked were those already having computers; the study population was therefore 61 secondary schools with computers. The schools were either teaching computer studies or were given computers by the government of Kenya. The total number of schools teaching computer studies in Kisumu County is 26; and, there are 35 schools in Kisumu County, (5 schools in each sub county) that received government grants in 2012 through Economic Stimulus Program (ESP) to procure computers and structure ICT centers for ICT integration in teaching all school subjects; such schools were therefore suitable for this particular study.

In this study, there were four categories of respondents that comprised the study population: 5,962 Form Four chemistry students from 61 public schools with computers; 61 school principals; 122 chemistry teachers, each school had averagely two chemistry teachers, though some schools   had one while others had three or four depending on the number of streams. For the sake of government policy on ICT integration, 7 Sub-county Quality Assurance and Standards Officers (SQASO) from seven Sub-counties in Kisumu County also formed part of the study population.

3.5 Sampling Techniques and Sample Size

This section dealt with sampling techniques and sample size.

3.5.1 Sampling Techniques

The research employed stratified random sampling and saturated sampling techniques to select a sample of schools for the study from the study population. Authorities like Gay (1987), Oso and Onen (2009), Krathwohl (2003) support the view that where a population is large and showing divergent characteristics it is possible to use more than one sampling technique in one study to enable the researcher to get a more representative sample. Stratified random sampling technique was used to ensure that among schools with computers, every category or type of school was covered in the sample. Stratified random sampling technique is a technique that identifies sub-groups in the population and selects from the desired sub-group to form a sample Sankaran (2003), (Gay, 1987). From the schools, principals, teachers and Form Four chemistry students were selected for the study. The Form Four students were selected randomly from each sampled schools to minimize relevance of bias. Stratified sampling technique was therefore used to ensure that the study population was divided into specific homogenous strata and that the desired stratum was represented in the sample in a proportion equivalent to its size in the population. This was to ensure that the specific sub-group characteristic was represented in the sample, thus raising the external validity of the study.

Saturated sampling technique was used to select Sub-County Quality Assurance and Standards Officers (SCQASO) from each Sub-county in Kisumu County for the study. Saturated sampling technique is a technique that is used to determine when there is adequate data for a study to develop a robust and valid understanding of the study phenomenon (Bloor & Wood, 2006), (Legard, Keegan & Ward, 2003). Saturated sampling technique ensured that the data was rich and insightful. The perspective of SCQASO was therefore critical for the implementation of ICT integration in public schools.

3.5.2 Sample Size

The sampling unit was the personnel: principals, teachers and students in the schools and the SCQASOs who were to be the respondents. Thayer (2020) defines sampling unit as singular value within a sample database. In this study the individual respondents were regarded as the sampling unit. Principals, teachers and students from 10% of the schools in the study (accessible) population were used for pilot study and were therefore not included in the final study. Since 10%, or an equivalent of six schools, six school principals and twelve chemistry teachers were used for pilot study; therefore, the study population from which the sampled was drawn was 55 schools, 55 school principals and 110 chemistry teachers. From these, 31% was drawn to make a sample of 17 schools; from which a sample of 17 school principals and 39 Chemistry teachers representing 31% and 35.5 % of the study population of principals and chemistry teachers respectively were purposively picked from the sampled schools. According to Mugenda and Mugenda (2008), a sample of 10% - 40% of the study population is enough for a descriptive survey. The 17 secondary schools that made 31% of the population were sampled from the seven sub-counties in Kisumu County on the criteria of the total number of public secondary schools (referred to as regular schools, see Appendix O) available in each sub-county at the time of data collection: Kisumu Central and Kisumu East with 13 and 15 public schools respectively, each had one school sampled, Kisumu West with 36 public schools, Muhoroni with 34 public schools, Nyakach with 53 public schools, Nyando with 42 public schools and Seme with 35 public schools each had 3 school selected randomly for the study. This gave a total of 17 schools sampled for the study. Saturated sampling technique was used to select 6 SCQASOs for the study, as one was used for pilot study.

Krejcie and Morgan (1970) Table for a large population (see Appendix P) was used to determine the representative sample size of 364 Form Four chemistry students from the 17 study schools sampled. The minimum number of participants needed for a pilot study was calculated at 24 participants as described by Kisser and Wassmer (1996). They applied the Upper Confidence Limit (UCL) approach to the sample size calculation and found that the pilot trial sample size of 20 – 40 would minimize the overall sample size for the main study sample. Consequently therefore, this study used 24 Form Four Chemistry students for the pilot study and left a sample of 340 Form Four Chemistry students for the main study. A sample frame showing how each group was sampled is provided in Table 4.

Table 4: Sample Frame

Subjects	Population size	Sample size	Percentage
SCQASO	7	6	85
Chemistry teachers	122	39	32
School principals	61	17	28
Chemistry students	5,962	340	6
Total	6,097	402	6.59

Table 4 gives the sample frame for the research. It gives a summary how each group of respondents was sampled for the study and gives the sum of the sample.

3.6 Data Collection Instruments

The data collection instruments for this study were four Questionnaires, an Observation schedule and a Lesson Observation checklist. The researcher was mainly concerned with views, opinions and perspectives of stakeholders in education; such information could best be collected through use of Questionnaires (Kothari, 2004; Krathwohl, 2003). At the school level the Questionnaires were administered to: School Principals; Chemistry Teachers and Chemistry Students. The fourth Questionnaire was administered to the Sub-County Quality Assurance and Standards Officers (SCQASO) in charge of ICT programs in each of the sub-counties within Kisumu County. In order to confirm the information gathered through the Questionnaires and to establish practical extent of available resources for integration, the researcher used an Observation schedule. And, in order to assess the level of integration of ICT, the researcher attended some ICT integrated lessons and used a Lesson Observation Checklist to do assessment during the sessions.

3.6.1 Questionnaire for School Principals

Questionnaire for School Principals (Appendix A-Questionnaire for School Principals) was used to gather information on the preparedness of school on implementation of ICT integration in teaching chemistry in Kisumu County. It had 46 closed-ended items divided into six sub-sections. The items were rated for availability of ICT equipment on a Likert-like scale from 1= none, 2= (1-3), 3= (4-7), 4= (8-9) to 5= (10 or more), that required the administrators to state the extent of availability of ICT software and hardware resources at their institutions. The instrument also required the administrators to give their perspectives as stake holders on the influence of ICT integration on KCSE chemistry performance in terms of deviation from the mean on a

Likert-like scale as follows: 1= Negative, 2=Not at all, 3=To some extent, 4= To a great extent and 5=To a very great extent. The main reason for structuring the questionnaire for school principals was to gather first hand information on the state of ICT integration at their institutions.

3.6.2 Questionnaire for Chemistry Teachers

Questionnaire for Chemistry Teachers (Appendix B-Questionnaire for Chemistry Teachers) had 58 closed-ended items divided into nine sub-sections. The instrument captured chemistry teachers' opinion on the of implementation of ICT integration in secondary schools; structured on a five point Likert-like scale that required them to state the extent of availability of hardware and software from 1= none, 2= (1-3), 3= (4-7), 4= (8-9) to 5= (10 or more). They were also assessed on the frequency they used ICT software in pedagogy on a Likert-like scale rated from 1=Not at all, 2=Termly, 3=Monthly, 4=Weekly, 5=Daily. The teachers were also assessed on the extent they performed certain tasks with ICT on a Likert-like scale from 1=Never, 2=Seldom, 3= occasionally, 4= frequently, 5= Always. The teachers were also required to give their perspectives on the influence of ICT integration on KCSE chemistry performance in terms of deviation from the mean on a 5 point likert-like scale. The main purpose of including teachers was to serve as a verification tool as the main implementers of the curriculum.

3.6.3 Questionnaire for Chemistry Students

Questionnaire for Chemistry Students (Appendix C-Questionnaire for Chemistry Students) had 45 closed-ended items divided into 3 sections. It was mainly used to verify the information given by both the teachers and the principals. They are the actual beneficiaries of ICT implementation; therefore their responses gave true picture of the state of the matter on the ground. The instrument required the students to verify the availability of ICT resources in their school, confirm the extent and frequency of use of available software in teaching and learning chemistry and specific topics that have been taught using ICT in their institutions. The items on the students' questionnaire were structured on a five point Likert scale inquiring on the availability of ICT resources, extent of use of the resources in teaching and learning chemistry, the extent of use of software in teaching and learning chemistry.

3.6.4 Questionnaire for Sub-County Quality Assurance and Standards Officers

Questionnaire for Sub-County Quality Assurance and Standards Officers (Appendix F-Questionnaire for Sub-County Quality Assurance and Standards

Officers) was administered to SCQASO in charge of ICT at the Sub-County. The items were used to solicit expert opinion on the official ministry position of implementation of ICT integration in secondary schools in Kenya. The instrument had 15 items; it had open-ended structured questions that required the officers to state the government position on training of teacher on ICT integration, the level of availability of ICT resources in public secondary schools. It also sought to determine the officers' official expectations of the extent of use of ICT resources by different personnel at schools. Besides the above, the instruments sought to determine from the officers the official government policy on capacity building of teachers with regard to ICT skills in pedagogy and their own perspectives as stake holders on the influence of ICT integration on KCSE chemistry performance.

This type of questioning was in line with what Kothari (2004) states that appropriate form of questions depend on the nature of the information sought, the sampled respondents and the kind of analysis intended. The line of questioning in this instrument was expected to address the research questions of the study appropriately. Since the structured instrument was in form of a questionnaire, it was administered thus; as questionnaires and not interviews. The researcher felt no need for face to face interview because there were no prompts and no scope for follow up questions to investigate responses which warranted more depth and detail. Because of the busy schedule of the officers who work in the field, the structure allowed the instruments to be administered quickly as in-depth was not required.

3.6.5 Observation Schedule

Observation Schedule (Appendix D-Observation schedule) is a tool essential in science used for observation to collect and record data. According to Kothari (2004) this method of data collection is very useful in extensive enquiries and can lead to fairly reliable results; he states further that non-response in such a tool is very low because it is filled by the researcher who is able to get answers to all questions. In this study the observation schedules enabled the researcher see and record what was available rather than rely on only what was stated as available in the questionnaires for principals and teachers in specific schools.

The schedule was used to make systematic check of the existing ICT resources in each of the study schools. This tool provided the researcher with reliable record of results from physical check on existing equipment, materials and personnel that aided users of ICT to integrate the subject. This physical examination assisted the researcher to have first-hand information on the availability of the resources in the study schools. The schedule was also used to collect archival data in form of KCSE examination records. The researcher used the results to record the study schools' overall performance in chemistry examinations for the period 2011 to 2017.

3.6.6 Lesson observation checklist

Lesson observation checklist (Appendix E- Lesson Observation Checklist for ICT Integrated Lesson) was used in this study to bridge the gap between what the respondents gave as what they do in ICT integration and what actually happens in the classrooms. In this respect the researcher was mainly concerned with teacher preparedness in the planning for and the actual ICT integration process in the classroom; such information could mainly be collected by lesson observation technique. As Oso and Onen (2011) explain, observation enables the researcher record information as it occurs and also notices unusual aspects. The lesson observation checklist used in the present study was based on a tool developed by Ottevanger (2001) in a context similar to the present study. The instrument was modified for the purpose of this study.

3.7 Reliability and Validity of the Instruments

Reliability measures the degree to which a particular measuring procedure gives similar results over a number of repeated trials while validity is the degree to which the empirical measure or several measures of the concept, accurately measures the concept. Orodho (2004) points out that content validity is a non-statistical method relying on a panel of specialists to assess the relevance of the content used in an instrument. Reliability of the instruments was established through piloting, while content validity of the instruments was established by experts' advice at the School of Education, Maseno University.

3.7.1 Pilot Study

Piloting of the structured instruments (questionnaires for students, chemistry teachers, school principals, quality assurance officers, the observation schedule and the lesson observation checklist) formed the first phase of this study. This was done in 10% of the population that were not involved in the final study. Piloting was done in six secondary schools that had computers; consequently six school principals and twelve chemistry teachers and twenty one Form Four students were selected for the exercise. The observation schedule, the ICT integrated lesson checklist and the Questionnaires were piloted on the same schools and personnel. The students'

Questionnaire was piloted on twenty one Chemistry students, three from three schools and four from three of the schools earmarked for the pilot study. The questionnaire for SCQASO was piloted at one sub-county, and involved one SCQASO who did not also form part of the final study.

Piloting was necessary because it helped the researcher to enhance the validity and reliability of the research instruments (Mugenda & Mugenda, 2008). Piloting also helped identify problems which respondents encountered in answering questions from the questionnaires hence served to confirm that the instruments exhaustively covered the required content. This enabled the researcher to revise the instruments based on the results of the pilot study in consultation with the supervisors.

The instruments were then re-tested on the same group of students, teachers, school administrators and SCQASO after two weeks and their responses were then analyzed manually by comparing their first and second set of responses in order to establish if the instruments attained a threshold reliability index. This was in an effort to ensure a satisfactory level of instruments functionality. After piloting the researcher made corrections on the items and discussed the instruments further with supervisors and designed the final version of all the instruments around the themes identified for the research.

3.7.2 Reliability

Reliability of the instruments was established through piloting by test-retest method. The pilot study that was done in 10% of the population was a means of ascertaining reliability of the tools. Reliability threshold index was realized at $\alpha = .70$ for the Questionnaires for the principals, $\alpha = .72$ for Questionnaire for Chemistry teachers and $\alpha = .71$ for Questionnaire for Chemistry students (see appendices A, B and C) using Chronbach's alpha, which were considered reliable. Findings were used to adjust the instruments so as to remove deficiencies and ambiguities. The observation schedule was tested at the same pilot schools. The results of the computer coefficients of reliability for the instruments led the researcher to conclude on the reliability of the instruments.

This study enhanced reliability during piloting of the instruments from different people in the same setting. With regards to reliability authorities such as Gall and Borg (2006) and Ferrell, *et al.* (2007) who subscribe to the view that researches originate from a variety of backgrounds and have different interests and inclinations. They argue therefore that reliability could be viewed in terms of comprehensiveness of data; the relationship between what is recorded as data and what actually occurred in the setting under study. Ferrell *et al.* (2007) further explains that reliability is enhanced by triangulation, whereby the same facts are elicited from different people in the same setting. This particular study enhanced reliability of results by comparing questionnaire results from the principals, chemistry teachers, students, observation schedule and lesson observation checklist in the same schools thereby enhancing the reliability of the results through triangulation.

3.6.3 Validity

This study relied on experts' advice at the School of Education, Maseno University to verify content validity of the tools. The experts ascertained against the objectives of the study to ensure that the instruments' contents measured what they were supposed to measure and therefore gauged their validity. All the questionnaires and the observation checklist were subjected to this treatment. The experts were supervisors and lecturers from the School of Education at Maseno University. The validity was first established by recommendation from the experts from the School of Education, Maseno University, who gave their views on the comprehensibility, relevance and clarity of the set items in the instruments to be used. The feedback was incorporated in the final instruments that were used in this research after the piloting.

3.8 Data Collection Procedures

Before any investigation into research problem was undertaken, the researcher obtained a consent letter from the School of Graduate Studies (SGS) at Maseno University after registering to research in secondary schools in Kisumu County. Clearance and permission was sought and thereafter approval obtained from Maseno University Ethics Review Committee (MUERC); this was in an effort to ensure that ethical issues in the research were adequately addressed (Appendix M).

After clearance from MUERC the researcher proceeded to seek clearance and obtain a permit from the ministry of Education, Science and Technology in an effort to be authorized to conduct research in secondary schools in Kisumu County (Appendix N). After the permit from the ministry had been obtained the researcher wrote an informative letter of introduction addressed to the Principals, Chemistry Teachers and Chemistry students of the sampled schools and SCQASO (appendix G, H, I and J) concerning the intended visit, issuing questionnaires and filling observation checklists. Arrangements were made with the principals for the suitable dates of meeting with the teachers and students to discuss ethical issues and in order to sign consent forms to participate in the research.

Once this was established, both quantitative and qualitative data were collected from 17 school principals, 34 chemistry teachers, 340 chemistry students and 6 SQASO during eight months of data collection using questionnaires and observation checklists. The researcher collected the data personally because precision was a requirement, and, as Kothari (2004) puts it; precision is an important factor to be considered at the time of selecting the method of collection of data; furthermore, self administration of instruments ensures that data was received by the researcher first hand. In this study precision in the observation checklists and lesson observation checklists was of paramount importance hence required the researcher to collect the data in person.

At the schools there was the handing over and completion of the questionnaires and completing the observation check list schedule. The principals, chemistry teachers and students completed the questionnaires. During the visit to each school the Director of studies was requested to take the researcher around the school's ICT center as then, observation checklist was completed and carried away. The questionnaires to SQASO were delivered to their respective offices after introductions had been conducted; the officers were allowed time to complete the questionnaires after which they were picked and compiled by the researcher. Later the researcher went back to the schools to conduct lesson observation of ICT integrated lessons for teachers who had indicated that they integrate ICT in their teaching and learning. The quantitative data was analyzed by descriptive statistics and presented in the form of tables, while qualitative data was analyzed by content analysis technique and presented in prose.

3.9 Data Analysis

Data was analyzed using descriptive statistics for quantitative data and thematically for qualitative data. Quantitative data obtained from the questionnaires were first serialized, coded then keyed into Statistical Package for Social Sciences (SPSS) computer program to generate frequency distributions, percentages, measures of central tendencies and variance, and presented in form of Tables; as Lochmiller and Lester (2017) explain, descriptive statistics involve identifying patterns, trends or frequencies in data.

The qualitative data from SCQASO on ICT school project were organized into categories and sub-categories as they emerged from the data, analyzed thematically and reported in prose. Qualitative data from the Lesson Observation checklist were coded then keyed into SPSS computer program to generate frequency distributions, percentages, measures of central tendencies and variance, and presented in form of Tables. The information was further analyzed thematically and reported in prose in the specific objectives. Kothari (2004); Lochmiller and Lester (2017) recommend such an approach when dealing with qualitative data.

A summary describing the data analysis process was presented in Table 5.

Table 5:   Summary Table describing Data Analysis Process

Objective number	Variables		Method of data
	Independent	Dependent	Analysis
1	ICT resources availability level	Learner performance	Descriptive statistics Thematic analysis
2	Teacher ICT literacy level	Learner performance	Descriptive statistics
3	Teacher ICT integration level	Learner performance	Descriptive statistics thematic analysis
4	Stake holders perspectives on ICT integration	Learner performance	Descriptive statistics Thematic analysis

Table 5 gives a summary of the dependent and independent variables in the various objectives and methods of data analysis process to be used in the study.

3.10 Ethical Considerations and Consenting Process

Preceding data collection process, respondents or study participants were taken through a recruitment process that culminated into counseling sessions in line with Belmont principles (Ryan, 1979) that guide the conduct of research using human participants. The issues of confidentiality, anonymity, informed consent and plagiarism, including consent for persons who are minors were addressed.   A detailed consent form was developed for respondents to read and consent before signing (see Appendix K). Learners were given consent forms to read and understand. Only the learners who signed the consent forms took part in the study. The consenting process contained three elements: information, comprehension and voluntariness. According to the Belmont principles (Bethesda, 1978) that guide conduct of research using human participants', justice and respect for person's demand that subjects enter into the research voluntarily and with adequate information. Voluntary response and honesty was clearly stated to the respondents.

Confidentiality and anonymity, according to the Belmont principles, urges the professional to keep a secret arises from the fact that harm will almost certainly follow if the information is revealed (Ryan, 1979). This study endeavored to maintain high standards of confidentiality and anonymity as professionally recommended. Issues like personal addresses, names of individual school principals, teachers and individual schools were concealed and codes were used instead in order to maintain confidentiality. Further, for confidentiality, anonymity and convenience the schools were coded using numbers from 1 to 17 and school type identified only as mixed day for sub-county schools, extra-county and county boys or girls schools, special school or national school for boys or girls, the same code for each school was maintained throughout the study.

Special provision was also made when comprehension of the research content would be limited by conditions of immaturity like learners who are below the age of eighteen years. According to Belmont principles, even for such learners who are considered minors the researchers should acknowledge their wishes and give them the required respect to choose the extent they are able, whether or not to participate in the research (Ryan, 1979) their objection should be honored. Because of respect the researchers in this study have for such underage learners, permission for their participation in the research was sought from their teachers who were expected to protect the subjects from harm.

Plagiarism is defined as presenting another person's work as one's own work. Presentation includes copying or reproducing it without the acknowledgement of the source (Narhe, 2005). It involves copying of phrases, clauses, sentences, and paragraphs or longer extracts from published or unpublished work (including from the Internet) that exceeds the boundaries of the legitimate cooperation. This study respects and values other peoples work and as such references were appropriately cited and acknowledged.

3.10.1 Storage and Protection of Data

It was the obligation of the researcher to protect the data obtained from the respondents. According to Kombo and Tromp (2006), protection of data is crucial in any research because it includes the regulations for processing personal information such as records kept in papers and data held in form of software in computers. In this context data obtained from this research was stored in two separate places for security, protection and accuracy before, during and after data analysis.

According to Baker (2018) while secure storage media will protect data when it is not being analyzed, it is also important to follow practices that keep data secure while it is being analyzed. To avoid incidental loss, there was double entry for the data both with password only known to the researcher and data analyst. One set of data was stored in a hard drive with a backup in a flash disc in the office with a password, and another set was keyed in and stored in the researcher's laptop at home, also with a password. This also ensured accuracy because the researcher would run comparison checks for the two sets of data to look at where there was no similarity. This was because one cannot make the same mistake twice. For every data there was a password for access to the database. The objective was to deny access for unauthorized persons. The researcher and data analyst were the only persons allowed entry into the database because of the requirement of a password. This ensured protection and security of the data as explained by Baker (2018).

Further security and protection to the data was censured by having a backup in a cloud programming safe in a secondary storage media or secondary hard disc that was stored separately from the main computer. This ensured that in case of theft of the computer or laptop due burglary the research team would still have the same data to fall back to.

CHAPTER FOUR

DATA ANALYSIS, INTERPRETATION AND DISCUSSION

4.1 Overview

This study investigated ICT preparedness, integration in education and stake holders' perspectives of ICT integration on chemistry performance among public secondary schools in Kisumu County, Kenya; using descriptive survey and sequential exploratory design. ICT integration was characterized by availability of ICT resources, extent of integration of ICT, teachers' capacity to integrate ICT and stakeholders perceptions of ICT integration on performance of chemistry students at KCSE examinations. This was in the light of the need to devise methods of teaching chemistry integrated with ICT to make its abstract concepts easier to the level of understanding of secondary school students to improve their performance. The data collected was analyzed using descriptive statistics and thematic analysis. This chapter presents the results of the analyses.

For systematic presentation and analysis of data, this chapter has been organized based on research objectives as follows:

i. Assessing the availability of ICT resources for teaching Chemistry in public secondary schools in Kisumu County.

ii. Establishing Chemistry teachers' capacity to integrate ICT in pedagogy.

iii. Establishing the extent of integration of ICT in teaching and learning Chemistry in public secondary schools in Kisumu County.

iv. Determining stakeholders' perspectives on the influence of ICT integration on KCSE Chemistry performance.

4.2 Demographic characteristics of the respondents

Demographic information providing data regarding research participants is presented in this section of the report. This was necessary to enable the researcher determine whether the individuals who participated in the study were a representative sample of the study population for generalization purposes. This was done in order to provide a clear background of the subsequent study findings and to establish if the study attained the required threshold. Data was mainly collected from the respondents through questionnaires. The questionnaires were administered to 6 SQASO, 17 school principals, 39 chemistry teachers and 340 chemistry students. The actual distribution and response received from respondents was recorded in Table 6.

Table 6: Questionnaire Completion Rate

Respondents	Questionnaires Administered	Questionnaires Fully completed	Percentage
DQASO	6	4	66.67
Principals	17	12	70.59
Students	340	340	100.00
Chemistry teachers	39	31	79.48
Total	402	387	96.26

Table 6 gives a summary of questionnaire completion rate. It summarizes how the questionnaires administered to different respondents were distributed, completed and returned. Overall 96% response rate was achieved on four sets of Questionnaires administered, this was considered commendable. Berg (2004) states that response rate of 70% and above is good. This is also an indication of a positive response rate; which could be attributed to the fact that the researcher administered the questionnaires personally. Use of personal administration of questionnaires was important for the researcher to get first hand information from the respondents; and, the observation checklist and schedules were personally completed with maximum precision; it was also a cost effective way to quickly collect the required information from a large number of people in a relatively short time.

Other research instruments employed included observation checklists in the study schools and secondary data collected from the KNEC on Kisumu County and its neighboring counties on performance in KCSE 2013 to 2016 (KNEC, 2017). The research instruments solicited background information on ICT qualification levels and work experience of the respondents such as the SQASO, Principals, teachers and study schools' ICT staff. This background information of key respondents was imperative to confirm if the research reached the targeted audience and if the research captured the information it sought effectively. The first demographic characteristic analyzed was academic and ICT qualification of the personnel, more so, chemistry teachers and ICT staff, presented in Table 7.

Table 7: Academic Qualification in ICT

Qualification in ICT	Number of personnel	Percentage
Post-Graduate diploma with IT	3	6.98
Under graduate degree with IT	14	32.56
Diploma with IT	10	23.25
Certificate in IT	14	32.56
None	2	4.65
Total	43	100.00

Table 7 gives a summary of demographic characteristics of the respondents. It specifically summarizes the academic qualifications of all the personnel that were involved in the study and responded to this particular question. Table 6 indicates that the respondents, who responded to this question, had attained minimum academic qualification in ICT; in addition to their professional qualifications, respondents who had attended certificate courses accounted for 32.56%. This was deduced to mean that over 95.35% of the respondents had adequate training in IT and hence had basic skills to integrate ICT in interaction with students.

The second demographic characteristic was work experience in administration among the principals of study schools in years as presented in Table 8.

Table 8: Work Experience of the Principals of schools

Number of years	Number of Principals	Percentage
0-5	7	41.18
6-10	4	23.53
More than 10	6	35.29
TOTAL	17	100.00

Table 8 summarizes the work experience of principals of the study schools. Among the principals sampled, 58.82 % have six or more years of administrative experience. It can therefore be concluded that majority of the principals have been in the study schools long enough to have experienced ICT integration taking off in their respective schools.

4.3 Availability of ICT Resources for Teaching Chemistry in public secondary schools in Kisumu County

The first objective of this study was to assess the availability of ICT resources in teaching chemistry in public secondary schools in Kisumu County. To achieve this objective principal, teachers and chemistry students were asked to state the quantity of equipment available at the ICT center; their responses were compared with the figures in the Observation schedule.

In response to this objective, data from principals, teachers, students and the observation schedule were analyzed by looking at the following per school:

i. Availability of infrastructure.

ii. Availability of computers and functional internet.

iii. Availability of ICT hardware for integration.

iv. Availability of ICT storage devices and content.

Data on this objective was analyzed under the research question "What is the extent of availability of ICT resources for teaching Chemistry in public secondary schools in Kisumu County?" The results were presented in Tables and prose.

4.3.1 Availability of Infrastructure per School

The objective in this sub-section focused on establishing the availability of ICT center or computer laboratory as had been indicated in the Observation schedule. The results for availability of an ICT center or computer laboratory were analyzed and data were coded according to the responses. For confidentiality the schools were coded using numbers from 1 to 17; the codes adopted for numbering the schools is maintained throughout the analysis.

An overview of the availability of computer laboratory or ICT centers in the study schools indicates that all the study schools have an ICT center except one, school number 15 whose ICT center was burgled, equipment stolen and have not been replaced. The availability of an ICT center or computer lab was deduced to mean that teachers in the schools can use the equipment to leverage or maximize their lessons through integration in lessons to present ideas dynamically by using a range of media.

4.3.2 Availability of Functional Computers against Form Four Student Enrollment and Internet Connectivity at ICT Center

An examination of the availability of functional computers in the ICT centers or computer laboratory was an important factor to inquire. The number of functional computers and those connected to the Internet was indicated in the respondents' questionnaires. This was confirmed by the researcher's observation checklist. The checklist results were analyzed and data on availability of Internet were recorded according to the responses and, this was compared with the mean chemistry performance per school and presented in Table 9.

Table 9: Functional Computers, Form Four Student Enrollment and Internet Connectivity per School

School	School Type	Functional Computers	Student enrollment (Form 4)	Internet Enabled Computers	Internet Connectivity	Mean grade
1	County girls	24	113	0	None	4.207
2	Special	12	36	1	Limited	3.277
3	Ex county boys	13	350	0	None	7.472
4	County boys	26	250	0	None	5.446
5	National girls	12	250	1	Available	7.472
6	Sub county	8	158	0	None	4.489
7	National boys	38	329	1	Available	9.383
8	County boys	23	160	0	None	3.894
9	Ex county boys	11	134	0	None	5.762
10	Sub county	11	64	1	Limited	3.652
11	Ex county boys	13	255	1	Limited	6.141
12	County girls	20	50	1	Limited	4.146
13	Ex county girls	18	180	1	Limited	4.262
14	Sub county	10	22	0	None	2.923
15	Sub county	0	72	0	None	3.326
16	Sub county	15	121	0	None	3.889
17	Sub county	9	93	0	None	3.751
	TOTAL	263	2637	7		

Table 9 summarizes the Form Four students' enrolment per school, the sum of functional computers and Internet connectivity at the ICT centers for the study schools. The researcher wanted to identify further, by inquiring, apart from the number of functional computers at the ICT centers, also, the number of computers that were Internet enabled and further, the availability of internet connectivity. It was observed that though majority of schools have 10 or more functional computers, the level of internet connectivity observed is very low; whereas seven schools have at least one computer at the ICT center enabled to the Internet; only two schools provide unlimited Internet services to their institutions. One national girl's school in Kisumu city collaborates with equity bank to provide Internet to their ICT center. Another national boys' school in the outskirts of Kisumu city collaborates with the government owned Telecom router that directs the Internet from the provider to various switches in the school; this resulted in the provision of unlimited Internet to their ICT center. Both the institutions utilized Wi-Fi mode of sharing Internet. Table 9 also indicates that five schools that have limited Internet available at their institutions utilize services of local Internet providers.

The rest of the schools have either limited or lacked internet connectivity completely. The remaining ten schools: Extra County, County and Sub County have not installed or subscribed to any internet service provider at all. It turned out that the national boys and girls schools with the highest mean scores of 9.383/12 and 7.472/12 respectively have unlimited Internet available at their ICT centers; while the sub-county school with the lowest mean score of 2.923/12 have no Internet connectivity at all at their ICT center. The study therefore established that Internet connectivity is not considered a priority in schools with ICT centers in Kisumu County as only the highly funded national schools have installed it.

The number of available functional computers per school was a matter of concern to the researcher, considering that a whole school population cannot adequately rely on computers in the computer lab for ICT integration in all subjects. The total number of computers among the study schools summed up to 263 against Form Four only student enrollment of 2,637. On analysis this gives an average student to computer ratio of 10 students per computer for the Form Four classes alone. This implies that for a teacher to take a class to the computer lab or ICT center for integration is an uphill task as it would put too much pressure on the computers at the center; that would render the machines prone to frequent damage hence non-functional.

Because of this challenge, the SCQASO were asked to give their opinion on the targeted number of students: computer ratio they desired to achieve at their sub- counties, this information was compared to their perceptions of the actual achieved ratio at the schools from research outcome. It was hoped that the targets could have been in line with the government target, as the policy document is still silent on the national target; but when asked, the quality assurance officers recommended an average ideal ratio of 5:1 students per computer. Sub-county wise the SCQASO were asked to indicate their perceptions on actual student: computer ratio achieved so far; their responses indicate that Kisumu East Sub-county had achieved a ratio of 49:1, Kisumu Central Sub-county had achieved a ratio of 63:1 students per computer respectively, and this could be because of urbanization in these Sub-counties. Muhoroni Sub-county achieved a ratio of 20:1 students per computer. Only Seme and Nyando Sub-counties had achieved a close to targeted ratio of 3:1 and 5:1 students per computer respectively because of their low populations, according to SCQASO. Their asserted results are illustrated in Table 10.

Table 10: Actual student: Computer Ratio per sub-county

Sub-county	Student to Computer Ratio
Kisumu East	49:1
Kisumu Central	63:1
Kisumu West	3:1
Seme	3:1
Muhoroni	20:1

Table 10 gives a summary of the student to computer ratio as it exists on the ground in the study schools according to SCQASO estimates. The finding in Table 10 reveals that a current status of an average of approximately 16 students per computer exists in majority of the study schools in the county. If this revelation is to stand test of time, then it implies that there is still need for the schools to acquire additional functional computers to meet the large student population especially in Kisumu East and Kisumu Central, urban areas, where the ratio stands at 49:1 and 63:1 students per computer respectively. This revelation is however considered outrageous compared to the actual findings at the study schools as revealed from the analysis in Table 9 that only reflects the population of Form Four students. The estimates by SCQASO contrast with national ICT policy document which states that in Kenya, even in schools that do have computers, the student: computer ratio is 150: 1 against a ratio of 15: 1 in the developed world and 45: 1 in Kenyan universities (MoE 2012). The findings in the current study compare unfavorably to the recommended universal learner to machine ratio of 3:1 (Mburu & Chemwa 2007), or to Britain where the ratios were 8:1 in primary and 5:1 in secondary schools in the year 2006 (Education Insight, 2006).

Recent study by Kiptalam (2011) confirmed the African and Kenyan situation and avers access to ICT facilities is a major challenge facing most African countries, with a ratio of 150:1 students per computer, against a ratio of 15:1 students per computer in developed countries; stating further that whereas results indicate that ICT has penetrated many sectors including banking, transportation, communications and medical services, the Kenyan education system seems to lag behind. A report by National Commission for Science, Technology and Innovation (NACOSTI, 2013) indicates that computer use in Kenyan classrooms is still in its infant phases; hence the extent of integration is also expected to be at the same levels.

This means that schools in Kenya still fall below expectation over a decade and a half later; this should give the government a desire to provide more machines for the Kenyan child in terms of availability of ICT equipment for integration of ICT in education. The situation also paints a lot of gloom for the level of ICT integration in Kenyan education because there are very many schools in Kenya where a computer has never been seen or used (Wabuyele, 2006). The very high learner: computer ratio indicated in schools in Kisumu County implies minimum access of the computers by individual students at the ICT centers, hence individualized ICT based learning sessions are seldom in the schools. Even though the situation is gloomy for Kenyan schools in terms of finances, the school administrators need to look for ways, in conjunction with the ministry of education, of equipping their institutions with computers to bridge the global digital divide.

4.3.3 Availability of variety ICT hardware for integration

It was part of the objective of this study to identify the variety of available hardware devises used for integration in order to establish the extent of their availability. To meet part of this objective there was need to ascertain the type and quantity of hardware devises for teaching and learning chemistry per study school. The researcher used the Observation schedule to record the existing hardware devices in the schools. The research findings revealed that ICT hardware devises found were available in different varieties and quantities at the ICT centers or computer labs per school; but averagely six varieties of hardware were identified in the study schools: Telephone or mailing devices, Video players, DVD and CD players, TV sets, Lap tops or tablets and LCD projectors. The varieties of hardware available in each school were therefore averaged out of the six types listed and converted to percentage as presented in Table 11.

Table 11: ICT Hardware Variety Available for Integration in Study Schools

School	Telephone or Mailing devices	Video	DVD and CD players	TV sets	Laptops or tablets	LCD projector	Average variety of Hardware (%)
1	0	1	1	3	1	0	67
2	0	0	0	1	0	0	17
3	0	0	1	1	0	0	33
4	0	1	6	1	0	1	66
5	1	0	1	1	1	0	50
6	0	0	0	0	2	1	33
7	0	0	1	1	1	0	50
8	0	0	1	2	1	0	50
9	0	0	0	1	1	0	33
10	0	1	1	1	1	0	67
11	0	0	1	2	0	0	33
12	0	0	0	1	1	1	50
13	1	1	1	1	0	0	67
14	0	0	1	0	0	0	17
15	0	0	1	1	1	0	50
16	0	0	3	1	1	0	67
17	0	0	1	1	1	0	50

Table 11 presents a summary of the hardware varieties that are available in the study schools. The Table indicates that among the study schools, DVDs and CD players are the most commonly available devises in the study schools; this could be because inbuilt CD player devices come with computers. While Television (TV) sets are the second most common devise in the study schools; with a sum of 19 TV sets in fifteen schools. Apart from the staffroom, students' common rooms are also equipped with the TV device for information, news and entertainment purposes, but none is available in the classrooms for use in teaching and learning. Occasionally, a class can access the TV by moving to the location or site of the TV to view videos for educational purposes.

Nonetheless, there is need to have a TV set in a classroom for educational purposes; this is because there are channels and websites for playing films, television and radio for schools that are licensed and are part of a course for instruction under the copyright act. According to the site this includes playing a film from a central point into a number of classrooms. Also, teachers can copy off-air television programs and broadcasts of previously broadcast free-to-air programs available on the broadcaster's website (Betz, 2011) to play to students as part of a course of instruction. This important process would only be possible if there are enough television sets in a school, to be distributed to classrooms for educational purposes other than just entertainment and news. The researcher in the present study opines the presence of a Television set in classrooms would imply that learners attach educational functions to a TV set other than just for entertainment and information even when at home; more so during quarantines when school functions were suspended due to a pandemic caused by the Covid-19 virus that was witnessed in the whole world. Under such circumstances, learners would be familiar with and would access educational channels and websites to continue learning at home when schools close.

Laptops are the third most common devise listed by the study schools according to Table 11, adding up to 12 devises in 11 institutions. The findings from the study reveal that most of the laptops available are personally owned by teachers. On being prodded further on what part their laptops played in ICT integration in their schools the teachers response was that occasionally they are used in classrooms for PowerPoint presentations or animations when need arises, but were mostly for personal use and research. It was therefore concluded that teachers had shown initiative to own lap tops even when not provided by their institution and use them for integration on few occasions, in line with the government desire to provide ICT integration in education. It is worthy to note here that it is only in developing countries like Kenya where laptop computers are considered as requirement only by teachers and probably university students for personal use, especially for research purposes; but not for high school students.

In the developed world there are programs for improving home access for high school students to own laptop computers for extended classroom-in-the-bedroom sessions; and more than 85% of pupils have their own laptops, and parents pay as low as $ 6 per parent per pupil over 3 years, and the funds used to provide laptop computers for the next generation of students (Cohen & Hill, 2001). If parents cannot or will not contribute, spare laptops are available for students. The implication of this to the current study is that it is possible for high school students to own computer lap tops. Though Kenya does not have any program or a recurrent budget for individualized laptops for secondary school students in public institutions, success may be achieved by institutions or counties adopting the soft mode of payment by borrowing a leaf from the developed world laptop program for high school students; this would increase the level of ICT usage by the learners for example school work at home sessions.

Video players are no longer considered as important gadgets in education systems and are slowly being replaced by DVD and CD players. Nonetheless this almost obsolete   gadget was found to be available in only 4 of the institutions of study. Its portability makes it easily accessible for classroom demonstrations when need arises.

Inquiries on the availability of LCD projector, the responses were summarized to show that the gadget was available in only 3 of the institutions in the study schools. This equipment is popular among teachers for PowerPoint presentations to students in class or at central places when demands arise. This finding is in contrast to the study by Eyup and Volkan (2017) whose findings indicated that Interactive White Boards are expected to replace classical boards; the study revealed that teachers make use of LCD panel interactive boards technologies frequently in their classes for educational activities despite some infrastructure problems and lack of software, stating that advantages in the use of interactive boards were more than the draw backs. Kenya is still lagging behind with this kind of technology, the high costs involved that goes with maintenance are not considered a priority in high school setup. Despite all these setbacks, the projector is an important tool for a school to own. It is possible that with financial assistance from educational stakeholders, these tools can be availed in more schools; and, probably the digital divide can be narrowed. The limited numbers of projectors in the study schools imply limited or lack of whole class presentations.

According to Table 11 gadgets for mailing services was the least available devise in the institutions studied; being available in only two of the institutions and used for bulk short messages (SMS) as a way of communicating to parents and teachers due to large enrollment in the institutions; hence not considered useful for integration. The presence of an array of ICT devises in some of the institutions was an indication of opportunities availed to educators and learners concerned; on the other hand very low figures in terms of availability of variety of equipment spells doom for the schools in Kisumu County as the digital divide widens with the developing countries. The finding in this section implies that schools have minimum investments in external storage devices and probably rely largely on internal storage in the computers which cannot be sufficient for each and every individual subjects' needs.

4.3.4 Availability of Storage Devices and their Content

Part of this objective was to establish the availability of ICT resources in the study schools. This was because ICT integration involves use of software for any meaningful learning to take place. This is in line with Becta (2005) who sought to suggest directions for the development of new content for e-learning programs with an objective to survey the impact of ICT in education and its progress in classroom instruction. To meet part of this objective there was need to establish the available software storage devises in the schools. The researcher used the principals' Questionnaires, chemistry teachers' Questionnaires and the observation schedule to record the existing storage devices in the study schools, the results indicated that only one school, specifically school number 2; a special school lacks any form of storage device completely. The result also shows that only one school, specifically the national girls' school number 5, had e-library in place. This implies that teachers and learners have no way of accessing digital content for reference. At the same time the findings also show that 94.11% of the schools each have one flash disc, DVDs and CD-ROM as storage devices; these are resource materials that are utilized by teachers but not students, hence limit interactions. According to Zimmer and Carpi (2003), storage systems consist of hard drives, Flash Memory, Floppy Disk, and Optical Disks. Any form of storage devise tailored to suit specific purposes would be expected to serve individual institutions.

4.3.5 Digital Content for Integration in Storage Devices in Study Schools

In order to establish the quality of available resources, the principals and the chemistry teachers in the study schools were asked to state the type of digital content available in the storage devices that could be used for ICT integration during teaching and learning processes in chemistry. Their responses were compared with the Observation schedule results, summed up and averaged out of the possible 11 types (converted into percentage) of software identified in the study schools. The software was identified as: e-

Revision questions, e-Past papers, e-Syllabus, e-Topical questions, animations, e-examinations, e-Teaching and learning materials (e-models), e-schemes of work, e-Library, Internet accessed materials and e-Topical notes. The results were recorded and presented in Table 12.

Table 12: Available Digital Content in Storage devises in study schools (n=17)

School	A	B	C	D	E	F	G	H	I	J	K	Total	Average %
1	1	1	1	0	0	0	0	0	0	0	0	3	27
2	0	0	0	0	0	0	0	0	0	0	0	0	0.0
3	1	0	0	0	0	0	0	0	0	0	0	1	9
4	1	1	1	0	1	0	0	0	0	0	0	4	36
5	1	1	0	0	0	1	0	1	1	0	1	6	55
6	1	1	0	0	0	1	0	0	0	0	0	3	27
7	1	0	1	0	1	0	1	0	0	0	1	5	45
8	0	0	0	0	0	0	1	0	0	0	1	2	18
9	1	0	1	0	0	0	0	0	0	0	0	2	18
10	1	0	0	0	0	0	1	0	0	0	0	2	18
11	1	1	0	0	0	0	0	0	0	0	0	3	27
12	1	0	1	1	0	0	0	0	0	0	0	3	27
13	0	1	0	0	1	1	0	0	0	0	1	4	36
14	0	0	1	0	0	0	0	0	0	0	0	1	9
15	0	0	1	0	0	1	0	0	0	0	0	2	18
16	1	0	0	0	1	0	0	0	0	0	0	2	18
17	0	0	1	0	0	0	1	0	0	0	1	3	27
Total	11	6	8	1	4	4	4	1	1	0	5		

*Key: A: e-Revision questions, B: e-Past papers, C: e-Syllabus, D: e-Topical questions, E: Animations, F: e-Examinations, G: e-Teaching and learning materials (e-models), H: e-schemes of work, I: e-Library, J: Internet accessed materials, K: e-Topical notes.

Table 12 presents the list of averaged digital content available in the storage devices in the study schools. The results show that four schools have over 36% of varieties of software identified as available by the study. Five schools have 27%, while other five schools have 18% of the variety of storage devices identified respectively. Two schools have less than 10% of the software while one school, the special school number 2 has no digital content in any external storage device at all.

E-revision questions, e-past papers, e-topical questions and e-examinations are digital content mainly used by teachers to retrieve materials for revision and evaluation of their students. If students can access computer storage devises, they would also retrieve materials for revision, especially in preparation for examinations. The varieties of revision materials are popular with many Kenyan teachers who believe that KNEC base their set questions on past papers. Moreover, many high school teachers tend to drill their students on answering questions available in past KCSE papers.

In Kenya, performance is pegged on examinations mean grade, therefore many teachers would prefer to access examinations materials readily in their local storage devices, and drill their students on how to answer the examination questions, rather than use the content for inquiry based learning with their students. Because e-examination materials are meant for evaluation, they are more of reference materials, not considered useful for integration in pedagogy. E-syllabus and e-schemes of work are mainly used by teachers to develop their own schemes of work and daily lesson plans; tailored to their specific classes.

E-topical notes mostly serve as resource for teachers to access major content for structuring their own lesson notes. Animations are mainly used for presentations to illustrate abstract content to learners. The models for teaching and learning are mainly used for structure and bonding of atoms and molecules, structures that can only be imagined are made visual. This way the abstract content of chemistry is made to appear simple. E-library and Internet accessed materials avail on-line resources mainly used by teachers for references and research; and by students doing inquiry based learning or project work.

From Table 12 it is evident that none of the study schools has invested in curriculum based software for teaching chemistry. This digital content was lacking completely and teachers relied on the various revision materials to access questions for revision, because revision is meant for evaluation this implies that actual integration is not practiced.

According to African Virtual University (Onwu & Ngamo ,2013) e-Learning project on ICT integration in education-option chemistry, the materials and equipment required for the module include computer software and data logging equipment in chemistry, portable ICT devises for modeling and simulation, worksheet, spreadsheet, database templates and graph drawing software, web-based resources: for interacting with appropriate teaching and learning chemistry materials on CD-ROMs, websites and interactive multimedia display boards and word processing facilities. The current research agrees that the items listed above are basic requirements for ICT integration to effectively pick up; the research findings therefore imply that very limited interactions can occur in the study schools with the software that were available, because they were neither adequate nor sufficient to support integration of ICT effectively.

The current study views the popular presence of e-revision questions as a move by institutions from accessing bulk hard copies of revision papers in form of mock papers previously common in the 1980s and 1990s, which were common before the digital age, to accessing soft copies of the same materials for revision which do not constitute integration. A parallel study by Kounenou *et al.* (2015) established that learners had high confidence in gathering data or getting support through synchronous online learning environments and multimedia, both learner-learner and learner-instructor interactions were significant predictors of student performance. The findings in the current study imply that in Kenyan high schools ICT and other media utilization is rarely by learner-learner interactions, the teacher interactions are pegged on search for information or content; which are in turn hardly displayed to the learners. The implication is that Kenyan learners are still more dependent on information given by their teachers; especially at high school level.

This finding is in contrast with the theory of Nilsen *et al.* (2016) from which the study was modeled. The model suggests that instructional quality determine student performance and specifically defined the relationship; the variable "teaching method by integrating ICT" to represent the stimuli, while "students' academic performance" to represent the response. In the current study the ICT equipment that would represent the stimuli was not sufficient to be utilized effectively for the theory to be effective. The theory could not apply to the study schools as the ICT enhanced environment was not availed to stimulate the required response from the learner.

These findings add to the body of literature surrounding traditional educational paradigm that still dominates education system, where the pupils' principal learning resource is the teacher; and the role of technology is to act as a carrier of learning subject matter, for exercise, for repetition and for feedback; and evaluation of pupils' progress is summative evaluation. This finding is in agreement with the views of Higgins and Moseley (2011), Law *et al.* (2008), Loweless *et al.* (2001) and Nachmias *et al.* (2004) who also express the same view. ICT in education is described by Mechlova and Malcik (2012); Jansen (2004) as a tool for construction and discovering science; a creator of context that supports learning by doing, a social medium supporting learning by communication and an intellectual partner of pupils that support learning by reflection. According to the researcher, evaluation of pupil's progress with relation to ICT includes evaluation of achievement tests, portfolios, formative evaluation, self-evaluation and evaluation of classmates.

The current researcher holds a similar view, when the teacher is the main resource; he or she also becomes the main source of assessment, narrowing the scope of teaching and learning.

The finding in the current study is in agreement with those of Karsenti *et al.* (2012) on Pan African research agenda who posited difficulty in the effectiveness of ICT into the learning process as statistics indicated success and challenges identified as Internet connectivity. This is confirmed further by the government of Kenya that recognizes this problem in its ICT strategy for education and training policy document (MoE, 2006), specifically, the policy document describes challenges of ICT in Kenyan secondary schools focuses on Internet connectivity and affordability (MoE, 2006). Report by Kiptalam (2011) posits lack of Internet or slow connectivity has affected most schools including high costs involved. This challenge is confirmed further by a public report of NEPAD e-schools demonstration project by Ferrell *et al.* (2007), who state that even where access to high speed connectivity is possible, high costs remain a barrier to access. This was established by the limited Internet connectivity experienced in majority of the study schools. The researcher therefore opines that the government should device ways of connecting schools to affordable Internet distribution services.

From findings in the current study, computers were not enough to be taken to classrooms although a number were available at the ICT centers in some of the schools; networking of the computers was not done in any of the study schools. Any form of integration would therefore be done by teachers maneuvering space for their classes at the ICT center or access presentation through Power Point using their laptops, both of which are very tedious exercises considering high enrollment in Kenyan classrooms; and therefore hardly ever done. This left the teachers helpless in terms of when and how to integrate ICT in their classrooms without computers. This is in agreement with the NEPAD e-school project, where according to Oracle (2005), each classroom was supposed to have a computer, served from ICT center for teachers to use in the NEPAD e-schools program. Furthermore, according to Ayere (2009) all the computers in NEPAD schools were also supposed to have been networked and connected through satellite from the African Computer Services Center to allow intra and external communications in the schools; this was never done. This finding implies that the study schools are not prepared in terms of access for the utilization of ICT resources including Internet for integration of ICT in teaching and learning chemistry to take place effectively.

Further implications of this finding is that learners in Kisumu County are not benefitting from stimulating learning environments created by ICT as theorized by Nilsen *et al.* (2016). A study by Luhamnya, Bakkabulindi and Muyinda (2017) in a review of theories on integration of ICT, recommended that all avenues to foster integration of ICT in teaching and learning should be explored as it brings about interactive learning environments and helps students deal with knowledge in active, self-directed and constructive ways. It is therefore imperative that school administrators in Kisumu County tackle the challenges and re-consider the issue of providing affordable unlimited Internet to their institutions to encourage research among teachers and innovativeness among learners. This can bring about the onset of e-learning, as a mode of alternative curriculum delivery when face to face learning between learners and their teachers is impossible caused by instances like what happened with Convid-19 pandemic globally. When learners are familiar with on-line type of learning they would have no problem accessing materials on their own; even in the absence of their teachers.

In summary the study established that the number of computers and other ICT equipment, resources and accessories including hardware and software associated with ICT integration are inadequate and limited in the study schools, available only at the ICT centers or computer laboratories; hence insufficient for meaningful integration. Implying that traditional educational paradigm still dominates education system in Kenya. Integration of ICT in teaching and learning chemistry is difficult to implement as the classes are large due to high enrollment. The study further established that Internet connectivity is not a priority among the study schools; available only in two national schools. The findings therefore reveal that the available ICT resources at the study schools are inadequate for teaching chemistry.

4.4 Chemistry Teachers' Capacity to Integrate ICT in Pedagogy

The second objective of this study was to establish chemistry teachers' capacity to integrate ICT in pedagogy. To achieve this objective, chemistry teachers in the study schools were asked to react to several statements intended to describe the status of their ICT literacy, their capacity to use digital software in teaching and learning and their capacity to perform pedagogical tasks with ICT.

In response to this objective, data from Chemistry teachers and SCQASO were analyzed by looking at chemistry teachers' capacity which was judged by:

1. Teachers' ICT literacy.
2. Teachers' capacity to perform ICT tasks.
3. Teachers' capacity to integrate ICT in pedagogy.
4. Teachers frequency of ICT integration

Data on this objective was analyzed under the research question "What is the capacity of chemistry teachers to integrate ICT in pedagogy?"

4.4.1 Chemistry Teacher's ICT Literacy

The first part of the research objective focused on the type of ICT literacy that chemistry teachers possessed. To explore on their ICT literacy, the teachers were asked if they had any training in ICT to gain skills that they could utilize in instruction during chemistry lessons. Literacy in ICT or training levels of teachers in ICT was further measured in terms of duration of training in ICT pursued by the teachers in addition to their professional qualifications: a degree certificate with Information Technology (IT), a diploma certificate with IT, a degree or a diploma and a certificate in IT that lasted up to one year, a degree or diploma and a certificate in IT that lasted for less than one year or None at all. The literacy level for the teachers in each of the 17 schools was determined; the outcome was summarized in Table 13.

Table 13: Chemistry Teachers ICT Literacy per school

School	Teachers' ICT literacy per school
1	5,3
2	5,4
3	5,4,4,2,2
4	5,4,4
5	5,4,2,2
6	5,2
7	5,3,2
8	5,2
9	5,4
10	5,2
11	4,3
12	5,3
13	4,2
14	5
15	5
16	5,3
17	4,2
Average	3.61

*Key: Teachers' ICT literacy: 5-Degree in Ed. with IT, 4-Diploma in Ed. with IT, 3-BEd/Dip with up to one year certificate course in IT, 2-BEd/Dip with Less than one year certificate course in IT, 1-BEd/Dip with None at all

Table 13 gives a summary of ICT literacy among chemistry teachers per school; and also the average ICT literacy for all chemistry teachers in the study schools. Inquiries revealed that for undergraduate courses, diploma and degree holders in Kenyan universities, at the turn of the millennium, all courses in education included a unit in IT. For practicing teachers, individual teachers funded own one year or less than one year certificate course in IT privately.

According to the findings, in fourteen schools all teachers have at least a degree certificate with IT while the other teachers either have diploma with IT or have under gone less than or up to one year course in IT. In the other three schools, teachers have either a diploma with IT or have undergone less than or up to one year a certificate course in IT; the schools are number 11, 13 and 17; an extra-county boys' school, an extra-county girls' school and a sub-county mixed school respectively. In the other twelve schools teachers have at least a degree certificate with IT with their professional qualification.

The average ICT literacy level for all the teachers in the study schools was determined and found to be 3.61, indicating that averagely teachers in the study schools have professional Diploma certificates in Education with IT. From the findings, teachers with similar ICT literacy in different types of schools posted different performance for their students depending on their school types and set traditions. Also, in some cases, teachers with better qualifications like a degree with IT posted poorer grades than teachers with less than one year course in IT because of the school type.

This finding add to a body of literature surrounding teachers' training on technology and use of technology in the classroom, suggesting that despite numerous plans to use technology in schools, teachers have received little training in the area of pedagogical use of the technology (Wachiuri, 2015; Mukuna,2013; Afshari & Shamah, 2009; Unwin, 2005; Hawkridge, 2002; Hawkins, 2002). When technology is introduced into teacher education programs, the emphasis is often on teaching about the technology, instead of teaching with technology; resulting in inadequate preparation to use the technology in pedagogy. Traditionally the need to learn to use technology is the primary drive for many teachers to acquire IT skills; after which assumption is made that the skill is sufficient. Even so, majority of research literature states that despite the frequent recommendation for teachers to have command on the use of word and data processing; the less frequently emphasized command of skills such as use of educational software and Internet brings a lot of challenges to practicing teachers (Usun,2009; Eurydice,2001). There is also research evidence to suggest that presently in some countries, the curriculum for each ICT in teacher education program, more practice based courses are included such as instructional technologies and material development (Alev, 2003). That this is lacking in teacher training in the Kenyan context was evident in the present research as all the teachers were IT literate; but were most probably not trained for pedagogical use of the technology.

Nonetheless, the findings indicate a response to the final report on the re-alignment of the education sector to the constitution that teachers are expected to put a deliberate effort to enhance their ICT literacy, skills and capacity in the area of ICT integration (MOE, 2012). According to the report, this will enhance their efficiency in service delivery and also empower them to play their rightful role in the envisaged knowledge economy. From the study, practicing teachers responded to this call by taking short certificate courses, this has given them the capacity to perform basic ICT tasks; but the skills to integrate ICT in teaching and learning go beyond basic skills of operating ICT equipment.

For Kenyan teachers, it appears that CEMASTEA training sessions for science and mathematics teachers is the only program standing out as means of In-servicing teachers on professional development on ICT integration; because the researcher was concerned with the ministry commitment on training this category of teachers. When this question was posed to the quality assurance officers as to whom the ministry uses to train practicing teachers on ICT integration, only one officer stated that CEMASTEA is mandated with professional development of all mathematics and science teachers, including ICT integration in pedagogy. The other officers fumbled with their answers, with one officer stating that there was none at all, while another stated that schools outsource for trainers. Another officer stated that teachers undergo self-training as other relevant resource persons may be used; this could be the true picture at the ground as most teachers in the study schools indicated to have taken own initiative to take less than one year course in learning computer packages. Another officer in answering this question stated that the ministry uses computer experts, without clarifying where the experts were sourced from. But one officer stated that the ministry uses ICT champions trained at the national level; but this was a one-time training that was done to jump start the implementation of ESP program; meaning that thereafter the ministry has not developed other modes of training practicing teachers on ICT integration.

In order to confirm the ministry position on teachers capacity to integrate ICT in pedagogy; it was important for the researcher to explore Kenyan government's commitment to capacity building among teachers to equip them with skills to integrate ICT in teaching and learning, inquiry was made through quality assurance officers (SCQASO) to establish what part the government plays on teacher literacy in the ICT integration process. The officers were asked among other questions, to state the people they liaise with at the school level in the monitoring and evaluation of ICT integration in high schools.

The SCQASO responses on whom they liaise with at school was summarized as follows: the quality assurance officers were almost unanimous that they liaise with the principal of the school (whom some referred to as administrators) and Heads of Departments (HODs). Only one officer stated liaison with HODs and teachers in schools. When asked to state the specific teachers they worked with at the school level on matters of ICT integration in pedagogy, half of the quality assurance officers stated that they worked with the computer studies teacher, computer champion or teacher in charge of ICT, but one officer stated that he worked mainly with science and mathematics teachers, while one officer stated that he worked with all teachers including the principal of the school, the HODs and learners. Since the principal of the school is the executive officer; it was concluded that the quality assurance officers liaised with the principal, and worked top-down with other teachers at the school level for the smooth running of ICT integration program. The ministry policy on quality assurance and standards expect the officers to liaise with the principal of the school who is in charge of implementation strategies in an institution (MOE, 2012). The SCQASO can only liaise with other teachers in a school through the principal to maintain protocol.

It was important to establish the part the ministry played in capacity building of teachers in ICT literacy, the quality assurance officers were therefore asked to state the kind of training the ministry gives to the teachers to encourage implementation of ICT integration according to the government policy document on integration of ICT in teaching and learning. One officer stated that there was no training offered at all on ICT integration, another officer stated that the ministry trains all science and mathematics teachers through CEMASTEA; another officer stated that the ministry trains teachers in basic skills in computers, yet another officer stated that the ministry trains teachers in ICT integration in teaching and learning, without stating the body contracted to do the job, but another officer stated that the teachers are trained in ICT integration on schemes of work by CEMASTEA. One officer stated that the teachers are trained on capacity building in their subject areas.

These diverse opinions by quality assurance officers is an indicator that the ministry does not have any specific program for capacity building or training of teachers on ICT literacy and integration; each officer gave his or her personal opinion from past experience; the researcher arrived at this conclusion because there was lack of uniformity in opinion among the quality assurance officers on official matters of education. The source of this mix-up could be in the implementation of training by CEMASTEA, it would be important to involve all stake holders in the training programs, such that administrators and the quality assurance officers are able to do follow up by monitoring and evaluation, which is currently lacking in the implementation strategies after training programs.

The other question posed to the quality assurance officers on teacher training was the duration or length of time training sessions for teachers took, two third of SQASOs stated that the training duration was seven days; one officer stated that the training duration was five days; whereas one officer did not respond to this question at all. Again it was clear the officers from the ministry were not unanimous in answering this question, meaning that there was no specific program on teacher training known to them. Therefore, when a question was posed as to who finances the training of teachers on ICT skills, seventy percent of the respondents stated that the Ministry of Education (MOEST) does the financing, but the rest of the respondents stated that schools finance the training sessions. It is not understood why some officers believe that schools can finance the training of the teachers because all the funds to schools originate from the ministry, except school feeding programs. Even the Center for Mathematics and Science Teachers (CEMASTEA) gets ministry funds to run their programs.

The quality assurance officers were asked which policy guidelines the ESP program followed in training of teachers from schools that benefitted from the government grants to procure computers. Again their responses were as diverse as their numbers, as one officer did not respond to this question, another officer stated that all secondary schools should be trained according to the government policy without specifically highlighting which policy he was referring to, while one officer stated that all teachers were expected to be ICT compliant without stating how they were to be facilitated. Another officer stated that in teachers' professional development, there should always be In-service training of teaching staff; but he did not clarify how it was programmed. On the other hand, one of the officers was of the opinion that the policy on ICT integration for Vision 2030 was the one in force for teacher training; while another officer was of the opinion that CEMASTEA module was the guideline used to train teachers. Once again these diverse opinions indicated lack of a single policy guideline by the Kenyan ministry of education in training of teachers as the officers were believed to be acting on past experience on responding to the above questions.

Despite the fact that all chemistry teachers at the study schools were IT literate and had indicated to perform some tasks in lesson preparation and in pedagogy with ICT, when the researcher visited each school a second time to carry out lesson observation of ICT integrated lessons, the findings were in contrast to what was indicated earlier in the Questionnaire for teachers. All the 17 schools were visited a second time in order to set up an appointment for lesson observation; this was successful in only 5 schools as follows: schools number 2, 4, 9, 12 and 14. In the other 12 of the study schools the appointments failed to take off as the chemistry teachers in each school were not in a position to carry out an ICT integrated lesson for varied reasons. For the researcher this was an implication that ICT integration does not take place in the schools as earlier indicated by the same teachers. The reasons for aborted lesson observation appointments were noted down as follows:

1. In school number 1 a county girls' school, the researcher arrived for appointment on two occasions, but the chemistry teacher, a female graduate teacher was not prepared on both occasions. On the second appointment date, the teacher confided in the researcher that personally she had never used ICT in teaching. She stated that occasionally there were some CDs in school on animations that had been shown to chemistry students in the past but presently could not be traced. She explained that chemistry teachers do not need ICT to teach chemistry, but carry out class practical or demonstration experiments, and that's all. In sciences generally, she stated that ICT integration may be done once in a very long time but it is not routine, the reason why the CDs could not be traced. Otherwise, she stated that ICT is common with computer studies where the subject is tested at KCSE.

2. In school number 3 an extra county boys' school, the researcher arrived for the appointment on three occasions, all of which aborted. On the last day the chemistry teacher confessed that ICT equipment are available in the computer lab but are mainly used by teachers who teach computer studies, who often integrate ICT in all their lessons, and occasionally by physics teachers, because there is a lot of digital content on electronic topics. The teacher confessed further that chemistry teachers fear issues that involve preparation, entailed in ICT integration, the teachers just want to go to class and teach. The teacher explained further that when

CEMASTEA team visits the school for monitoring and evaluation on the extent of ICT integration in the school, they are referred to the teacher in charge of ICT, because they do not want to get involved. Therefore, the teacher confessed that there is no integration of ICT that takes place in the school.

3. In school number six, a sub-county school, no appointment was issued; the chemistry teacher was frank with the researcher and declared that ICT integration is never done in that school mainly because the school does not have a hall and the classes are overcrowded. The school has a projector but due to lack of space, even in the overcrowded classrooms, ICT integration is not practicable.

4. School number 10 is a sub-county mixed school that has two projectors, but no appointment was fixed for lesson observation; the chemistry teacher declared before the principal and the researcher that chemistry teachers never integrate ICT, it is considered a waste of time by teachers in the department, hampering syllabus coverage; according to them ICT is left for computer studies and biology departments who have a lot of content for integration. She stated further that chemistry department had content on animations in the computer hard drive, but disappeared because of a virus and the computer crushed, leaving the department with no digital content.

5. In school number 13, an extra-county girls school, appointment failed to take place after 2 visits to the school, the chemistry teacher declared that ICT integration does not take place, neither is it routine in any subject except computer studies; science teachers are comfortable with practical in their subjects.

6. In school number 15, a sub-county mixed school, during a meeting between the researcher, the school principal and the chemistry teacher, information was given that ICT equipment was initially available in the school, but there was a burglary and everything was stolen, except the projector. Therefore there is no integration of ICT that takes place in the school.

7. School number five, a national girls' school was visited four times by the researcher in an effort to get an appointment to observe an ICT integrated chemistry lesson. All appointments failed to take place due to protocols. The chemistry teacher confided in the researcher that teachers in that school hardly Integrate ICT in teaching and learning, especially chemistry teachers. They believe it is a waste of time. He explained that what teachers need for learners to pass exams is content to use to teach,

because the syllabus is too wide; and, integrating ICT into lessons consumes a lot of time unnecessarily.

8. At schools number 7, 8, 11, 16 and 17 national boys schools, county boys', and sub-county mixed schools respectively, no successful appointment took place as the chemistry teachers failed to take up appointment for lesson observation, since they claimed that ICT integration was not applied in teaching chemistry; implying that ICT integration was not being practiced in the schools.

Sequential exploratory design of this study therefore, focused on observation of ICT integrated lessons that took place in only five different schools. A total of five teachers, one female and four male participated in the observations. ICT preparedness of the teachers to integrate ICT was inquired and established as follows: all the five teachers were graduates, three had graduated in their professional degrees in education with IT, one had a bachelor of education degree and had pursued a certificate in computer packages specializing in animations, while one teacher had a bachelor degree in science education and had pursued a diploma in IT. The teaching experience of the teachers varied from one year to seventeen years. The classes in which the lesson observations took place were large in terms of enrollment with an average of forty eight learners. On average only one lesson per teacher was observed.

The lesson observation checklist instrument contained a set of statements about the background of the teacher and different dimensions on the level of ICT integration in the classroom. The statements in the checklist were based on teacher preparedness for the lesson, extent of integration of ICT specifically the resources used, method of delivery and learner engagement with the resource. The scoring of the lesson observation checklist during classroom observations provided a picture of how the lesson progressed on a rating scale as follows: zero if the dimension was not applicable, applicable dimensions scored 1 to 5 on a scale from poor to excellent.

Observation of ICT integrated lesson was the main part of the sequential exploratory design. Five teachers participated in the observations in their chemistry lessons. The identities of the teachers were withheld and school number codes used: T2, T4, T9, T12 and T14 for teachers from school number 2, 4, 9, 12 and 14 respectively. Characteristics of the teachers and their schools are presented in Table 14.

Table 14: Lesson Observation: Teacher Preparedness

Teacher/school	T/2	T/4	T/9	T/12	T/14
School type	Special	County boys	Ex-county boys	County girls	Sub-county
Gender	Female	Male	Male	Male	Male
Teaching experience (years)	1	17	1	10	1
Academic qualification	BEd Sc	MSc	BSc	BEd Sc	BEd Sc
Computer literacy	Degree with IT	Degree with IT	Degree with IT	Certificate in IT(animations)	Diploma in IT
CEMASTEA in-set sessions	None	5 sessions	None	3 sessions	None
Other in-service courses	None	None	None	ESP	Starter computer packages

Table 14 indicates that all the five teachers have almost common professional background. One teacher has a Master of Science degree with seventeen years teaching experience and has attended five CEMSTEA In-service courses. The other teachers have Bachelor of Science degrees in education, with one year experience for three teachers and ten years experience for one teacher who has also attended three sessions of CEMASTEA in-set and an ESP training session. Three teachers studied IT with their degree, while one teacher pursued a diploma in IT and a certificate in computer package specializing in animations. The teachers who are new in the profession have not attended any CEMASTEA in-set courses for science teachers.

This profile implies that the five teachers have shown interest in ICT from basic training to in-service and additional courses in IT. This finding is in line with what Stuart (2015) in his study of ICT integrated models into middle and high school curriculum in the USA stated, in order to prepare students with skills and knowledge necessary for information society, ICT should be prepared at all levels and in all subject matter in chemistry curriculum appropriately. According to the researcher, to this end, teachers need to be prepared with skills and knowledge of academia, ICT and pedagogies of both for the integration. The implication to this study is that merely teaching IT should not be the goal of education; ICT should provide opportunities to learners to learn better in an enjoyable environment, and to teachers to enable their learners attain their potential.

4.4.2 Capacity of Chemistry Teachers to Perform ICT Tasks

Even after training to acquire literacy in the use of technology, it is important for teachers to embrace the technology by constantly applying its use in order to improve confidence. It is therefore imperative that teachers take self initiatives to adopt ICT to improve their proficiency hence performance for the new information age. The type of tasks the teachers were able to perform with ICT and the proficiency with which they performed such tasks during chemistry instruction was assessed.

4.4.2.1 Chemistry Teachers Capacity to Perform Administrative tasks with ICT

In order to explore the specific ICT proficiency of the Chemistry teachers, it was important to inquire on the tasks the teachers applied with ICT in administration and instruction and the extent they applied such tasks. To explore the specific ICT proficiency of chemistry teachers they were asked to state the frequency they applied technology to perform specific tasks that included tasks to: navigate tutorials, browse by searching for tutorials, perform administrative duties like compiling marks, creating worksheets, creating talk books, use ICT in lesson preparation and perform functions like animating diagrams, building web animation, downloading images, drawing charts inserting videos and pictures on to existing documents in lesson preparation for chemistry instructions. From the teachers' questionnaire, data on the responses of the teachers on their ICT proficiency were summarized in Table 15.

Table 15: Chemistry Teachers capacity to Perform ICT Tasks in Administration and Lesson Preparation (n =39)

Task	5	4	3	2	1	Mean	SD
			Frequency of Task Performance				
Navigating tutorials	10	10	0	9	10	3.03	1.61
Searching tutorial	15	8	0	5	11	3.28	1.73
Compiling marks	23	6	0	3	7	3.9	1.60
Creating work sheets	19	9	0	3	7	3.79	1.58
Creating talk book	2	10	3	7	17	2.31	1.40
Animating diagrams	2	12	2	10	13	2.49	1.37
Building web animation	1	7	0	2	29	1.69	1.28
Downloading images	6	12	4	6	11	2.90	1.50
Drawing charts	6	7	1	12	13	2.51	1.50
Inserting videos	4	11	4	11	9	2.74	1.37
Inserting pictures	4	7	3	11	13	2.42	1.41

*Key: 5-Always, 4-Frequently, 3 -Occasionally, 2 -Seldom, 1-Never.

Table 15 gives a summary of tasks that chemistry teachers can perform with ICT as gathered from the teachers' questionnaire. It is important to note here that these were opinion of teachers of chemistry and stood to be confirmed by lesson observation; also some teachers did not respond to some questions causing fluctuations in the tallying. Table 15 indicates that about 50% and 58% of chemistry teachers always or frequently browse by technology to navigate tutorials or find tutorials for particular teaching tasks respectively. The means for these tasks were 3.03 and 3.28 with standard deviation of below 2 indicating the respondents agree that these tasks are only occasionally performed.

Apart from teaching, teachers usually have administrative tasks which they must execute; the researcher needed to distinguish tasks which teachers execute with ICT while on duty. In order to explore the capacity of chemistry teachers to perform tasks in administration with ICT, they were asked to specifically state the frequency with which they perform several of their administrative tasks using ICT. It was important to explore specific administrative tasks that the teachers perform with ICT in order to establish their proficiency with technology. They were therefore asked to respond to specific tasks involving ICT that they perform in administration. The administrative tasks inquired of the teachers were compiling of marks, creating work sheets, creating talk book and building web animation using ICT. Their responses summarized in the Table 15 indicates that about 74% and 72% of the teachers always or frequently perform their administrative task of compiling marks and creating work sheets using ICT respectively. The mean for these tasks were 3.9 and 3.79 respectively with standard deviation of below 2, meaning that averagely teachers in the study schools are in almost perfect agreement that they frequently compile marks and create worksheets of their students using ICT.

The frequency of creating talk book and building web animations indicates that about 62%and 79% of the teachers in the study schools never perform the tasks respectively with ICT at all. The means for these tasks were 2.31 and 1.69 respectively with standard deviation of 1 each, indicating that the respondents are in agreement that such tasks are seldom performed in the study schools. This trend implies that for a modern society such as Kenya; there is need for a paradigm shift.

Teachers were required to state the frequency with which they perform several lesson preparation tasks using ICT this was in order to explore the ICT preparedness of chemistry teachers in lesson preparation. The teachers were given specific tasks in lesson preparation which they were asked to respond to in the teachers' questionnaire.

The teachers were asked to state the frequency with which they could: animate diagrams on PowerPoint, download images from the Internet, draw charts or insert videos into PowerPoint presentation or insert pictures into an existing document with ICT in their lesson preparation. Their responses indicate that about 56%, 44%, 64%, 49% and 62% of the teachers in the study schools seldom or never perform these tasks respectively using ICT at all. The mean for the tasks were 2.49, 2.9, 2.51, 2.74 and 2.42 respectively all with standard deviations of below 2, indicating a perfect agreement that teachers seldom perform these tasks during their lesson preparation with ICT.

The findings in this sub-section indicate that teachers in the study schools are more proficient with ICT in performing administrative duties. All the other tasks involving lesson preparation were seldom performed with ICT by over 73% of teachers in the study schools.

This finding indicate very low figures comparatively on lesson preparation tasks performed by teachers with ICT, and contrasts with Fouts' (2000) research who asserted familiarity with common software such as word, excel, PowerPoint, web animations, downloading images, drawing charts and tables and other basic operations are important skills for teachers to possess. Previous research has pointed to teachers' lack of pedagogical computer competence as the main barrier to their acceptance and adoption of ICT in developing countries (Pelgrum & Law, 2009). The findings in the current research support and extend the findings from previous research. The majority of respondents reported to having seldom or occasionally performed tasks in lesson preparation indicate little or no competence in handling most of the computer functions needed for pedagogy by educators. This finding supports the assumption that teachers with low level of computer competence usually have negative attitude towards computers (Barak, 2007). In Kisumu County uptake of the technology is still low, it may require the intervention of the government to reconsider the need to retrain practicing teachers on pedagogical ICT integration and develop curriculum based software for teaching specific subjects like Chemistry to make use of the technology tenable.

4.4.2.2 Teachers Proficiency to Integrate ICT in Pedagogy

In order to explore the chemistry teachers' proficiency to integrate ICT in pedagogy, it was important to inquire the frequency by which they were able to perform various ICT tasks in teaching and learning during instruction sessions. This was in line with the research by Onwu and Ngamo (2004) and Ngamo (2006) who asserted that teachers must have certain competencies and abilities in order to support student learning with ICT. This supports the theory of Mechlova and Malcik (2012) that key elements of modern paradigm with relations to integration of ICT are pupils learning resources; and these are whatever can be a resource: textbooks, books, magazines, audio and video recording, Internet, electronic encyclopedias, classmates, teachers and specialists out of school; while, teachers must use varied methods and forms of teaching and learning process: group work in small heterogeneous groups, project methods, experimentations, finding and synthesis of information, presentations and large variability of learning activities. On this end therefore, teachers were asked to state the frequency by which they were able to perform several chemistry teaching and learning sessions using ICT. Their responses were computed and combined such that all tasks involving teaching and learning gave overall view of their opinion on performance and summarized in Table 16.

Table 16: Teachers' Proficiency to Integrate ICT in Pedagogy (n=39)

Task	5	4	3	2	1	Mean	SD
		Frequency					
Use interactive white board	5	1	4	9	19	2.05	1.39
Conduct ICT class presentation	14	3	0	12	10	2.97	1.71
Conduct online class discussion	2	3	0	9	25	1.6	1.15
Create a quiz on PowerPoint	5	4	0	13	17	2.15	1.42
Structure digital lesson notes	10	7	0	8	14	2.77	1.69
Teach students to write a report	1	8	0	15	15	2.28	1.52

*Key: 5-Always, 4-Frequently, 3 -Occasionally, 2 -Seldom, 1-Never

Table 16 summarizes the frequency of task performance by chemistry teachers in the study schools to integrate ICT in pedagogy. It was necessary to establish specific teaching and learning tasks that the teachers could perform with ICT. They were therefore asked to respond to specific questions regarding their capacity to perform tasks specified. The teachers were asked to indicate the frequency by which they conduct lessons using interactive white board, conduct an ICT class presentation, conduct an online class discussion, create a quiz on PowerPoint, structure digital notes and teach students to write a report using ICT. Their responses summarized in Table 15 indicate that about 72%, 56%, 87%, 77%, 56% and 77% teachers respectively who seldom or never perform teaching and learning sessions using ICT tasks. The means for these tasks are 2.05, 2.97, 1.6, 2.15, 2.77 and 2.28 respectively with standard deviation of below 2 for all tasks, indicating that averagely teachers in the study schools are in almost perfect agreement that they seldom use ICT in these pedagogical tasks, except conducting ICT class presentation which is done occasionally. Kenyan teachers therefore need more exposure on this aspect. Either lack of time to set it up in the classrooms or lack of interest is possible explanations.

There is a change in the role of the teacher from an instructor to a constructor, facilitator, coach and creator of learning environment (Catalan, 2007). This is supported by the theory of Jansen (2004) that describes a teacher's role in ICT integration as facilitator, entrepreneur, coach and creator of authentic experience. The implication of this finding is that for most teachers in Kisumu County, the skills attained are basic skills of operating the technology but not suited for integration of ICT in teaching and learning. Among practicing teachers in the study schools the capacity to use ICT was attributed to the skills acquired through training on use of technological tools; but skills to integrate the technology in pedagogy have not been sufficiently acquired. This finding is in agreement with the views of Alazam *et al.* (2012) and Ruales (2012) that revealed a significant correlation between ICT skills and ICT integration in classroom as majority of teachers never used ICT in teaching and learning in Malaysian schools, and confirmed that only perceived skill level is a significant predictor of integration of ICT to instruction.

Lesson observation process in the field revealed that even those teachers who allowed the researcher to observe their ICT integrated lessons, there was hardly any planning as all the lessons were hypothetical and arranged solely for purposes of the observation. All the teachers were evaluated for preparedness for the lesson just before the start of each lesson. The average scores in the section of teacher preparedness in lesson preparation in schemes of work and lesson plan in the observation checklist were summarized in Table 17.

Table 17: Preparedness for ICT integrated lesson: Summary

Dimension	T/2	T/4	T/9	T/12	T/14
			Teacher / School		
Schemes of work	0	0	0	0	0
Lesson plan	0	0	0	0	0

*Key: Not Available = 0

Table 17 summarizes scores for schemes of work and lesson plans for teachers who participated in ICT lesson observation exercise. The scores are zero for all participants because no teacher observed indicated adequate preparedness for the ICT integrated lessons by not having developed technology integrated lessons in either of the tools required for teacher preparation. The absence of technology integrated lessons in schemes of work and lesson plans is an indication of routine integration exercise is not planned for. The absence of both tools implies that ICT integration in teaching and learning chemistry is not routine practice that takes place in any of the schools under normal circumstances; and that the classes observed were hypothetical for the purposes of satisfying the researcher's request.

This finding confirms that having ICT equipment and IT literate teachers in a school has not impacted on traditional educational paradigm and contrasts the stimulus-Response theory by Nilsen, *et al.* (2016). As the theorists have indicated, teacher quality enhanced by instructional quality by integrating ICT to represent stimuli has not impacted on the teaching practice of Kenyan teachers to elicit the response of improved student academic performance. Having formed an attitude towards traditional methods of teaching with their students performing as per their ability, participants in the present study have not gone out of their way to incorporate modern teaching paradigm to their traditional approaches.

As theorized by Burton, Moore and Magliaro (2004) where the curriculum is determinate strictly, consistent standards, curriculum, subjects and lessons. And, according to the theory, methods and forms of teaching and learning, individualized ways and rate of pupils, typical low variability of learning activities are the norm. For the current study, findings imply that unless there is a shift from traditional educational paradigm to modern educational paradigm, performance of learners will continue to follow school type rather than quality of the teachers or equipments the schools may have.

This finding reflect a contrast of what Alvardo (2011) wrote in his report on what best practices in ICT-enhanced school should be like: an integration of ICT in the curriculum reflected in the teachers lesson plan. The former researcher opines that this requires a school to formulate a comprehensive school technology policy; and development of ICT- enhanced content by designing technology enhanced E-learning and Tele-collaborative learning projects. The findings in the current research indicate that ICT integration in the context of Kisumu County public schools is not considered as a curriculum obligation process. Accordingly, teachers do not integrate ICT in pedagogy, but continue to use traditional methods of teaching.

This finding is in agreement with views on learning theory of Mechlova and Malcik (2012) that more than half of teachers of natural-science subjects do not have skills and multimedia tools to support the development of key competencies of pupils (problem solving, learn how to learn, social and personal competencies, competencies for working with digital technologies) at their disposal during teaching; and, they do not have resources for learners to use computer aided experiments in natural science subjects. According to the researchers, results show that 93% of teachers are not confident with their skills to integrate ICT and want to further educate themselves in the technological field. For the current study this implies that the government of Kenya through the ministry of education should put more emphasis on pre-service and in-service teacher training on pedagogical ICT integration in teaching and learning, and, put in place monitoring and evaluation strategies through quality assurance and standards department of the ministry to ensure compliance with ministry objectives.

Findings in the current study add to a body of literature surrounding teachers IT literacy and actual use the technology in various administrative tasks, but they hardly use the technology in pedagogy. Usun, (2009) argues that philosophical and sociological underpinnings of ICT in education are missing in all the countries of Europe. Perhaps this can be explained by what Jonassen (2004), asserted that ICT as a modern paradigm in education creates new demands on teachers, and are not accepted by all teachers. According to the theory, many teachers are resistant to this approach or refuse it entirely. This view is supported by a study by Rastogi and Malhotra (2013) that revealed that success of implementing curriculum with ICT in education depends greatly upon the attitudes of teachers and their willingness to embrace such technology with ICT knowledge and skills. Recommending that teachers should possess not only ICT knowledge and skills, but also have to develop and imbibe right attitudes towards ICT; because these have a marked influence in their readiness to utilize technology in their teaching. The implication of this finding is that classroom practices in Kisumu County may not be influenced by ICT integration unless there is a paradigm shift in the mode of training of teachers' on technological pedagogical content knowledge (Mishra & Koehler, 2011). For teacher trainers, it is important for the candidates to be exposed to the effective use of ICT in Teaching and learning (Sketee, 2006).

In summary, the findings in this section revealed that among the study schools, all chemistry teachers are IT literate; despite this, none of the teachers in their lesson plans integrate ICT in teaching the subject as revealed during lesson observation exercise. Performance of schools in KCSE chemistry did not match ICT literacy of their teachers, but tended to depend on the school type. The findings further revealed that over 72% of teachers in the study schools frequently perform administrative tasks of compiling marks and creating worksheets with ICT; but seldom perform tasks in lesson preparation with ICT. In pedagogy, over 72% of teachers seldom perform tasks with ICT; the only tasks frequently or always performed by teachers are class ICT presentations by 43.5% and structuring of lesson notes by 44% of teachers. Lesson observation revealed that teachers do not practice ICT integration and do not consider it as complementary to their curriculum delivery process.

4.5 The Extent of Integration of ICT in Teaching and Learning Chemistry in Public Secondary Schools in Kisumu County

The third objective of this study was to assess the extent of integration of ICT in teaching and learning chemistry in public secondary schools in Kisumu County. To achieve this objective the question was answered at two levels: by the students who experienced the process of integration in their classrooms; and, by the researcher attending some ICT integrated lessons to confirm integration.

In response to this objective data from Form Four chemistry students, chemistry teachers and lesson observation checklist were analyzed by looking at:

1. Chemistry teachers' extent of use of ICT in pedagogy
2. Extent chemistry topics are taught through ICT
3. Observed practice of ICT integration

Data on this objective was analyzed under the research question "what is the extent of integration of ICT in teaching and learning Chemistry in public secondary schools in Kisumu County?"

4.5.1 Chemistry Teachers' Extent of Use of Educational ICT Software in Pedagogy

The chemistry students through their questionnaire were asked to rate their experience on the extent of use of some ICT equipment identified for teaching and learning the subject with their teachers. This was in an effort by the researcher to establish the extent of integration of ICT in the study schools. It was therefore important to establish specifically from the learners the extent that their teachers used educational ICT software in teaching and learning chemistry in the classrooms in order to determine the use of such software on learner performance. The learners were therefore asked in their questionnaire to state the extent they have experienced the use of educational ICT software in teaching and learning in their chemistry lessons. The results were summarized in Table 18.

Table 18:   Extent Teachers Integrate Educational ICT software in Pedagogy

Software	Frequency of use of software					Mean	SD
	5	4	3	2	1		
Internet access	41	38	5	9	247	2.00	1.57
E-library	43	12	7	7	271	1.77	1.49
Video shows	24	23	15	47	231	2.18	1.30
Content rich	43	19	17	12	249	1.93	1.52
Content free	36	19	12	18	255	1.82	1.44
CD-ROM topical revision	28	34	25	33	219	2.01	1.41
Websites	32	25	14	10	259	1.81	1.43
CD-ROM encyclopedia	24	27	14	11	265	1.73	1.34
Microsoft excel	72	43	13	19	192	2.57	1.73
Microsoft office	74	43	12	21	189	2.59	1.74
T/L software	44	33	12	25	225	2.10	1.56
Training software	29	36	8	14	252	1.86	1.44
Audio shows	15	24	27	19	254	1.69	1.22
Simulated videos	15	20	21	16	267	1.60	1.18
Animated videos	19	25	24	4	268	1.69	1.28
Topical videos	29	32	28	36	214	2.03	1.41
Computer skills	63	43	17	31	185	2.51	1.67
Interactive learning	40	31	21	24	224	2.08	1.52
General revision	60	49	27	31	173	2.59	1.63

*Key: 5-Very Great Extent, 4-Great Extent, 3-Some Extent, 2- Very Little Extent, 1-Not at all

Table 18 gives a summary on the extent chemistry teachers integrate ICT software in pedagogy as responded to by chemistry students in their questionnaires. Table 18 reveals that in almost all aspects of teaching and learning over 70% of learners have never experienced the use of any ICT software in their chemistry lessons, with standard deviation of below 2 in all aspects, indicating an almost perfect agreement among student respondents. Specifically 72.64%, 79.70% and 71.17% of the student respondents have never experienced the use of Internet or Internet accessed materials, E-libraries or access to websites respectively during their chemistry lessons. The analysis on Table 18 also reveals a mean response of 2.00, 1.77 and 1.81 respectively on the use of this particular software, with standard deviation of below 2 in all aspects, indicating that averagely the learners are in almost perfect agreement to have experienced the use of Internet and Internet resourced materials as ICT educational software to very little extent. This mode is considered too high for lack of access to such important ICT equipment.

The major hurdle, specifically, in the access of Internet and E-libraries which are also resources from the Internet in high schools in Kenya could be the lack of connectivity and the cost of Internet data bundles; of which Kenya has got one of the most expensive rates globally; and which is not catered for in the government funding at secondary school level of education.

It is also of concern to note that majority of institutions of secondary education in Kisumu County having not invested in E-libraries which is considered as very affordable and useful software. It is much cheaper for an institution to install E-libraries in their computers hard drive and other internal or external storage systems than to construct a depository specifically built to contain books and other materials for reading and study (Atherton, 2011). It may be helpful for institutions to consider the option of E-libraries as a resource center for their teachers and learners. The situation needs a paradigm shift; teachers need to be encouraged by availing sufficient resources, to adopt new methods of teaching through discovery by research.

Other educational software illustrated by analysis on Table 18 reveals that 67.94%, 73.23% and 75% of learners have never experienced the use of video shows, content rich software and content free software respectively in their chemistry lessons. The means indicated by these responses are 2.18, 1.93 and 1.82 respectively with standard deviation of less than 1.5 indicating that averagely the learners in the study schools are in agreement that they are exposed to this software to "very little extent". Effects of video display have been researched; findings in the present study are in contrast to a study by

Yehudit *et al.* (2014) who established that the visual illustration captures the interest of the learners, asserting this may enhance their comprehension of the abstract concept and in effect improve their performance. Perhaps it is with this in mind that few institutions of secondary education in the study have to adopt video shows as part of curriculum delivery process.

Topical revision, topical videos software refers to revision questions and subject content arranged topically detailing both chronological and topical basis, while general revision provides overall revision questions. Table 18 indicates that 62.94%, 62.94% and 50.88% of the respondents respectively have not experienced the use of topical revision questions, topical videos subject content or general revision materials stored in CD-ROM in their chemistry lessons. The mean for this responses are 2.01, 2.03 and 2.59 respectively, with standard deviation of less than 2 indicating that averagely the respondents are in almost perfect agreement that they have experienced use these software to very little extent respectively. This could have been due to the culture of Kenyan learners accessing revision text books widely distributed in Kenya to aid all forms and types of revision hence the need for schools to stock e-revision materials, another reason could lack of required resources to enable learners' access these ICT resource when needed.

Use of CD-ROM encyclopedia showed a total of 77.94% of the student respondents have never used or experienced the use of CD-ROM encyclopedia in any of their chemistry lessons. The mean for this response is 1.73 with a standard deviation of 1.34 indicating an almost perfect agreement that the respondents averagely use the software to very little extent.

The integration of Microsoft Excel and Microsoft Office are the only ICT software that has been experienced for use by a total of 24.41% and 25.08% of the students respectively to a "Very great extent". It is worth noting that despite this revelation, there were still 56.47% and 55.58% of the respondents who have never experienced the use of this software in their chemistry lessons at all. The mean for these responses are 2.57 and 2.59 respectively with standard deviation of 1.73 and 1.74 respectively, which indicate an almost perfect agreement that averagely learners' use Microsoft excel to some extent.

This finding is in contrast with educators' overview from Microsoft (2017) on their report to unlock limitless learning for students. They aver every educator wants to see their students succeed in learning and in life. They advise educators on Microsoft education that provides the tools to create an inclusive classroom and personalize the learning for every student, enabling them to become more self-directed, confident learners, when student experiences are driving the core of Microsoft then the meaning is always personalized learning for every student and gives every student opportunity to shine with tools that help educators personalize learning for diverse classrooms and nature self directed confident learners.

Teaching and Learning software, training software, audio shows, simulated videos and animated videos are used to learn chemistry topics offline. A total of 66.17%, 74.11%, 74.70%, 78.52%, 78.82% and 62.94% of the respondents have never experienced the use of these software respectively in their chemistry lessons at all. The mean for these responses are 2.10, 1.86, 1.69, 1.60, 1.69 and 2.03 respectively, with standard deviation of less than 1.6 indicating an almost perfect agreement that averagely learners experience the use these software to very little extent. It is not understood why these important software were hardly utilized in the study schools. A simple explanation could be an overcrowded syllabus leaving very little space for comprehensive revision, or lack of time to dedicate to revision hence the utility of the software and ICT equipment that requires commitment and dedication; or inaccessibility of the ICT resources.

Software for computer skills enables learners gain expertise on the use of computers as a prerequisite to learning with computers; while interactive learning software enables learner-learner and teacher-learner interactions in discovery learning. A total of 54.41% and 65.88% of the learners respectively have never experienced the use of these software in their chemistry lessons at all; the mean for these responses are 2.51 and 2.08 respectively, with standard deviation of less than 1.7 indicating an almost perfect agreement on average use of the software to some extent and to very little extent respectively. One of the reasons for this scenario could be due to lack of the software in the institutions, lack of access to adequate infrastructure to use it or probably lack of proper planning for ICT integration and utilization of the available resources.

The responses from this analysis indicate a big percentage of respondents lack exposure to usage of the software caused by low levels of integration by teachers. From the findings the mode of the responses tends towards "Not at all" in using the various ICT software inquired, even in cases where the equipment is available it is inadequate. The implication to the current study is that there is need for the government of Kenya to put in place measures for sufficient supply of ICT integration resources, before it can be said to have any effect on learner performance.

4.5.2 Extent Chemistry Topics are taught through ICT Digital Content

To investigate further on the extent of teachers' ICT integration, the researcher used the students' questionnaire to identify the extent of teaching certain chemistry topics in the study schools integrated with ICT digital content.

The summary of students' response on the extent of use of ICT centers for teaching and learning specific topics integrated with ICT digital content in chemistry is shown in Table 19.

Table 19: Extent of Teaching certain Chemistry topics by Use of Digital Content (N=340)

Chemistry topic	5	4	3	2	1	Mean	SD
Spectrometer in titration at end points	20	13	16	9	282	1.54	1.20
Duck builder in structure and bonding	28	18	16	10	268	1.70	1.35
Computerized Molecular Modeling and Visualization	17	19	16	15	273	1.58	1.20
ICT based chemistry laboratory (virtual)	19	17	20	15	269	1.16	1.22
Hot potatoes for quizzes	18	11	27	8	276	1.56	1.18
Power Point Presentations	26	15	18	13	268	1.67	1.31
Live Internet sites	22	14	15	13	276	1.59	1.24
Multimedia	31	24	19	9	257	1.82	1.42
Movie clips in slides	21	19	18	19	263	1.66	1.26

*Key: 5-To Very Great Extent, 4-To Great Extent, 3- To Some Extent, 2- To Very Little Extent, 1-Not at all

Table 19 gives a summary of chemistry teachers' extent of use of digital content at ICT center for teaching and learning some topical aspects in chemistry as responded to by chemistry students. Table 19 indicates that in every aspect of teaching and learning using ICT, over 75.5% of the respondents, who were chemistry students in the study schools have never experienced teaching and learning their chemistry topics integrated with ICT. Specifically, out of the student respondents 82.92%, 78.82%, 80.29%, 79.11% and 81.17% respectively have never experienced: the use of spectrometer in titration at end points, the integration of duck builder in structure and bonding, learning ICT integrated computerized modeling and visualization; or visualized an integrated ICT based or virtual chemistry laboratory, or have been exposed to hot potatoes for quizzes using ICT respectively. The means for these responses are: 1.54, 1.70, 1.58, 1.166 and 1.56 respectively; with standard deviations of less than 1.35, indicating that learners in the study schools are in perfect agreement that they have experienced integration of these chemistry topics with ICT to very little extent.

Similarly, 78.82%, 81.17%, 75.58% and 77.35% respectively of the student respondents have never been exposed to: Power Point presentations, live internet sites, learning through multi-media and video shows or movie clip on slides respectively in their chemistry classrooms. The means for these responses are 1.67, 1.59, 1.82 and 1.66 respectively, with standard deviation of less than 1.42 showing that averagely the learners are in almost perfect agreement that they have been exposed to this form of ICT integration to very little extent.

The trend shown in the responses above indicate that ICT integration in the public secondary schools has not taken effect significantly and this should be of concern to education stake holders on the state of ICT integration in Kenyan schools. This could have been due to lack of time in the schedule of teachers, lack of adequate supply of ICT equipment or lack of timetabling of ICT center for other subjects to utilize it for integration; a close scrutiny of the time table in all the study schools indicated computer studies as the only subject allocated time on the time table to the ICT center. Another reason could be due to lack of curriculum based software for teaching and learning chemistry, leaving the teachers unsure of the approach to take.

This finding is in contrast with the views of Onwu and Ngamo (2016) whose study revealed that web-based technologies are considered to be widely used for educational purposes; asserting that the process of integrating ICT in education is hardly a simple and straight forward one, the transition from traditional teaching to ICT enhanced environment is not always obvious. He established that overlaps in the application of ICT tools in teaching always occur; stating that ICT is sometimes used in combination with non-ICT strategies such as shifts from text books based to web-based books or from

PowerPoint presentation in class to PowerPoint presentation via the Internet. Sometimes they both operate in parallel, in conjunction or interchangeably.

This contrast with findings in the present study revealing that ICT enhanced environment does not exist in Kenyan classrooms. The implications to the current study is that web-based books, E-library or PowerPoint presentations are terms unknown to many learners in Kenyan classrooms and traditional text books and chalkboard presentations are the norm. This implies that a paradigm shift is required to encourage Kenyan teachers to adopt modern teaching methods in cases where ICT equipment is available.

This situation was confirmed during lesson observation exercise; teachers at the study schools were evaluated for the extent of integration of ICT; whereby the researcher identified the dimensions of the integration, specific ICT resource and extent of its use during the lesson and scored on a rating scale. The results were summarized for teachers in five schools who participated in the lesson observation exercise and presented in table 20.

Table 20: Extent of integration of ICT: Summary of resource used per teacher

ICT	Integration dimensions	Teacher/School				
		T/2	T/4	T/9	T/12	T/14
1	Suitability and adequacy	3	2	3	3	2
2	Variety and originality	2	3	4	4	1
3	Relevancy	3	4	3	3	1
4	Utility and effectiveness	2	3	3	4	1
5	Quality of technological skills	2	4	4	5	2
6	Display	2	5	3	5	1

*Key 0=Not available, 1= poor, 2= Satisfactory, 3= Good, 4= V. good, 5= Excellent

Table 20 indicates the dimensions on the extent of integration in terms of resources of ICT used during lesson observation of five teachers in different study schools. The teachers' scores were averagely satisfactory in all dimensional aspects except teacher number T/14 whose scores were bordering poor performance in four out of six activities.

During the lesson the teachers were evaluated on the level of use of ICT resource focusing on the method of delivery. The activities were summarized in Table 21.

Table 21: Method of delivery: summary of delivery process per teacher

	Level of use of ICT resource	Teacher/ School				
		T/2	T/4	T/9	T/12	T/14
1	Teacher demonstration	2	3	2	4	2
2	Class experiment	0	0	0	0	0
3	Expository approach	1	3	2	2	2
4	Inquiry learning	0	0	0	3	0
5	Discussion	2	2	3	4	2
6	Project	0	0	0	0	0

*Key 0=Not available, 1= poor, 2= Satisfactory, 3= Good, 4= V. good, 5= Excellent

Table 21 indicates that all the observed lessons were teacher demonstrations expository approach, no class experiment and, with limited discussions. Learner involvement was limited except for the class of teacher number T/12. Class experiments and projects were lacking and inquiry based learning was noted only in the class of teacher number T/12.

During the course of the lesson, the teachers were evaluated on learner engagement with ICT resource focusing on learner activities as summarized on Table 22.

Table 22: Summary of learner activities during lesson observation per teacher

	Learner activities	Teacher/School				
		T/2	T/4	T/9	T/12	T/14
1	Attention drawn to resource	2	2	3	3	2
2	Observe teacher demonstration	2	3	2	4	2
3	Demonstrate concept with ICT	0	3	3	2	0
4	Experiment with ICT	0	0	3	2	0
5	Discuss observation from experiment	2	0	2	3	2
6	Make conclusions on observation	2	0	3	2	2

*Key 0=Not available, 1= poor, 2= Satisfactory, 3= Good, 4= V. good, 5= Excellent

Table 22 gives a summary of learner activities during lesson observation of ICT integrated lessons for each teacher. According to the results, most presentations were teacher centered with activities requiring learner input limited to observations and listening to teachers' input during or after the presentation.

Details of classroom observations for individual teachers are presented here as general descriptions of all lessons. For teacher number two (T/2) it was clear the teacher, with only one year experience in the profession, had challenges of presenting a video clip to a class of sixty students using a single smart phone device. The projector was fixed at the ICT center and the class was too large to fit at the center. The resource used was therefore completely inadequate and considered not appropriate as an ICT integration process, the presentation was rated as poor. Teacher number four (T/4) had seventeen years of teaching experience, a degree with IT and five sessions at CEMASTEA in-set, he was quite competent and confident in class presentation. Though it was not routine exercise, the teacher managed to download content from Internet and presented it to class. The display projected on a large screen was quite audible and visibility was in order. The overall performance was rated as satisfactory.

Teacher number nine (T/9) with only one year teaching experience showed confidence in the skill using ICT with his class; though he had only a laptop to display a scientific process of separation of mixtures by sublimation. Because of the small screen the learners' scrambled for closer positions to view the screen. The overall performance of the teacher was rated as quite average. Teacher number twelve (T/12) had a lot of confidence due to long experience of ten years coupled with additional training in computer packages. He engaged his learners who were enthusiastic to participate in a contest. The overall performance was rated as satisfactory. Teacher fourteen (T/14) displayed confidence, with only one year experience in the profession coupled with a diploma in IT. Use of a single laptop for display for a class of forty was the main weakness. The presentation was rated as quite average.

During visits to the study schools, the researcher had the opportunity to evaluate the practicality and effectiveness of ICT integration in chemistry classrooms in the study schools. From the observations and discussions, the presentations were hypothetical and real ICT integration is not practiced. The findings indicate that ICT integration in teaching and learning chemistry does not take place, and has no effect on learner performance in public secondary schools in Kisumu County.

This finding contrasts the theory of Nilsen *et al.* (2016) from which the study was modeled. The theorists postulate in their model that teaching methods by integrating ICT represents "stimuli" that would elicit a "response" in student behavior that would culminate in improved performance of the learners. That the impact of ICT as stimuli could not be felt in the study schools hypothesized in the theory because direct link is mandatory between amount of practice, massing and distribution of trials, the schedules of reinforcement and other variables acting as stimuli that strengthen the learner's response to take place. Teachers in the study schools did not integrate ICT in teaching and learning to an extent that would elicit a response of improved performance from the learners.

That teachers are not keen in ICT integration in education had long been established by Jones (2002) from analysis of survey data collected in 2001 in the study in metropolitan Melbourne in forty six primary school classrooms over a four week period, the findings indicated very low level of ICT use. Every classroom surveyed had access to computers, but their use could be best described as being occasional. The data collected indicated that more than 90% of the supervising teachers in the survey used their classroom computers with students once per week or less for four week teaching practicum. The implication of this finding is that learners in Kisumu County will not be able to reap full benefits of ICT if teachers are not involved in determining effective ICT integration strategies in pedagogy.

Higher levels of skills and knowledge would produce higher levels of technology integration had been established by Kounenou (2015); Higgins and Moseley (2011); and, Afshri and Shamah (2009); who hypothesized that teachers belief in their computer competence was the greatest predictor of their use of computers in the classroom. The researchers point out that a teacher should develop competence based on the educational goals to be accomplished with ICT. This view contrasts with the findings in the current study, where all chemistry teachers are IT literate but fail to integrate ICT in their pedagogical process, and view ICT as a waste of time and blame the overloaded syllabus, or perhaps the Kenyan teachers have competence in computer use but lack competence in the use of the technology in teaching as the hidden motive behind the attitude indicated.

This finding adds to literature surrounding obstacles to the integration of ICT in the teaching process suggesting that the frequent problem has been the insufficient number of computers. This finding is similar to the findings of the study conducted by Pelgrum (2001) who points out that apparently most countries have not succeeded in realizing sufficient facilities to keep teachers up to date with new technologies to integrate these in teaching and learning. In his study Pelgrum also shows other non-material obstacles that most teachers have found difficulty in integrating ICT in instruction: scheduling enough computer time for students, insufficient teacher time, and lack of supervisory and technical staff. Other researchers share the view that the attitude of teachers using computers was an important factor; Pelgrum and Plomp (1991) and Abdelrahman *et al.* (2013) point that out as the key of success for the implementation and integration in instructional purposes. In Abdelrahman *et al.* (2013) study, the interview used showed that most teachers' attitude towards using computers was negative. Pelgrum (1993) indicated that teachers with positive attitude towards using computers were more likely to use the technology more frequently and intensively in their teaching subjects. The interview in the former research yielded that respondent teachers asserted difficulty to integrate ICT in teaching and learning.

In summary, the study revealed that over 70% of learners in the study schools have not experienced use of any ICT software in their chemistry lessons The study further revealed that over 75.5% of learners have never experienced teaching and learning of any chemistry topics integrated with ICT; including live Internet sites, virtual labs, or multimedia. Lesson observations of ICT integrated chemistry lessons at the study schools failed to take off in 70% of the study schools. In 30% of the study schools, hypothetical ICT-integrated classes were organized for the researcher to carry out the observation exercise, indicating that routine ICT integration is not practiced and does not take place in any of the study schools in Kisumu County, and by no means has any effect on performance.

4.6 Stakeholders' Perceptions of ICT Integration on KCSE Chemistry Performance

The fourth objective of this study was to assess stakeholders' perceptions on the influence of ICT integration on KCSE chemistry performance. To achieve this objective, school principals and chemistry teachers in the study schools were asked to react to several statements intended to describe their perceptions of ICT integration on KCSE chemistry performance; the chemistry KCSE performance of the study schools was retrieved from the observation schedule. In response to this objective, data from principals, chemistry teachers and SCQASO, and observation schedule were analyzed by looking at:

1. Performance trends in KCSE chemistry mean grades for the study schools.

2. Chemistry mean grade and deviation from the mean per school according to chemistry teachers

4. Chemistry teachers' perspectives on frequency of integration of software in teaching and learning and performance in the subject

5. Chemistry teachers' perspectives on mean usage of ICT and performance.

Data on this objective was analyzed under the research question "what are the stake holders' perspectives of ICT integration on KCSE chemistry performance?

4.6.1 Performance Trends in Study Schools from 2011 to 2017 in KCSE Chemistry

Examinations are the most commonly used means for evaluating the progress or lack of it among students. The examination administered by Kenya National Examinations Council (KNEC) is used as a means of externally evaluating students. As Shiundu and Omulando (1992) put it, examinations tend to dictate the curriculum instead of following it; learners assess education in terms of success in examinations. Teachers, too, recognize importance of examinations in the lives of individual learners and as a result relate their teaching to the examination system. It was important to establish teachers' perspectives on ICT integration with regards to the performance of their students at KCSE examinations, in order to establish their beliefs in learning theory of ICT as a stimulus in instructional environments and performance as the response.

Performance Trends between 2011 and 2017 in Chemistry subject as examined in KCSE examinations were investigated in the study schools and obtained through data collected using the observation checklist. It was necessary to obtain the performance trends of the schools in the examination to establish if the trends could have been dependent on, but not limited to ICT integration, category (type) of the schools, traditions established by schools or any other attributes in the population not considered in the study; for this reason trend analysis was used to gauge performance of individual schools but not to rank the study schools. The results were presented in Table 23.

Table 23: Performance Trends in Study Schools from 2011 to 2017 in Chemistry at KCSE

Type of School	2011	2012	2013	2014	2015	2016	2017	Mean
1.County girls	3.493	3.524	4.157	4.761	4.363	4.045	5.104	4.207
2. Special	3.375	3.500	3.110	3.419	2.838	2.980	3.720	3.277
3. Extra-county boys	8.721	6.261	6.618	7.066	8.209	7.782	7.645	7.422
4.County boys	5.210	4.661	5.300	5.860	6.520	5.290	5.283	5.446
5.National girls	7.218	6.278	6.946	7.605	8.209	7.463	8.585	7.472
6.Sub-cunty mixed					5.255	4.197	4.016	4.489
7.National boys	9.397	9.266	8.375	8.680	10.247	9.819	9.899	9.383
8. County boys	2.770	3.670	4.086	3.240	3.420	4.330	5.744	3.894
9.Extra-county boys	5.534	4.619	5.403	5.615	6.391	6.391	6.379	5.762
10. Sub-county mixed	3.490		3.410	3.360	4.010	3.720	3.920	3.652
11.Extra-county boys	6.424	6.425	6.416	5.484	6.662	5.775	5.802	6.141
12.County girls	3.481	3.375	3.187	3.659	5.000	3.821	6.500	4.146
13.Extra-county girls	3.310	4.190	4.320	4.458	4.893	4.160	4.500	4.262
14.Sub-county mixed	2.835	2.625	2.500	2.470	4.137	2.592	3.304	2.923
15.Sub-county mixed	2.558	2.800	3.255	4.706	4.501	2.649	2.813	3.326
16.Sub-county mixed	3.430	3.130	3.383	4.155	4.798	3.961	4.365	3.889
17.Sub-county mixed	2.770	2.670	4.086	3.240	3.420	4.330	5.744	3.751
Mean	4.354	3.940	4.385	4.575	5.463	4.90	5.489	4.845

Table 23 gives the trends of performance in the study schools from 2011 to 2017 and finally gives an overall mean for each school for the seven year period, a county mean for each year and for the seven year period. The trend in some of the schools lack consistency as the mean grades in chemistry tended to fluctuate from as low as less than 3.5/12 for more than 10 sub-county schools in the year 2011. The trend indicated for some of the schools was fluctuating improvement shown up to a maximum in 2015 (except for school number two, a special school) but dropped in 2016 and 2017 to give a mean grade of below 4.5/12 in sub-county schools. This is an exception of national school for boys with a mean of 9.383/12, and national school for girls with a mean of 7.472/12. Other schools that did not show much change in their chemistry mean grade are extra-county schools for boys with a mean of 5.762/12, and 7.472/12 respectively. Others include one county school for boys with a mean of 6.141/12 and one county mixed day school with a mean of 5.446/12. These are the schools whose mean grades remained more or less constant from 2011 to 2017. The trends indicated by the study schools do not show variation in teacher quality or instructional quality that would have warranted a response of improved performance among learners. The trends show ordinary performance in evaluation; the performance could have been influenced by school type more than any other stimuli.

This finding is in contrast with the study by Wastiau *et al.* (2017) on the use of ICT in education during a survey of secondary schools in Europe that established that students who were taught with ICT had better academic performance with ICT instructional package, recommending use of ICT for teaching and learning in secondary schools. This implies that in the current study, the presence of ICT equipment in the study schools did not impact on learner performance because they were not integrated in teaching and learning.

4.6.2 Chemistry mean grades and deviation from the mean per school according to chemistry teachers

The chemistry teachers were specifically asked to state the extent ICT integration had affected chemistry mean grades by estimating deviations from previous means (ranging from negative deviation to positive deviation up to +3) in their schools for seven years of ICT integration. Their responses were compared to the school chemistry mean grade and frequency of use of ICT by the same teachers per school, the summary was presented in Table 24.

Table 24: KCSE Chemistry mean Grade per school, perceived deviation from previous mean and Teachers Frequency of ICT Integration (N=17)

School	chemistry KCSE mean	Deviation from previous school mean	Teachers mean frequency of ICT usage	SD
1	4.207	3	2	1.20
2	3.277	3	1	0.45
3	7.472	3	2	1.54
4	5.446	4	4	0.69
5	7.472	4	3	0.90
6	4.489	3	2	1.59
7	9.383	5	4	0.98
8	3.894	2	2	1.72
9	5.762	4	5	0.92
10	3.652	3	3	1.58
11	6.141	3	3	1.39
12	4.146	4	5	1.01
13	4.262	3	2	0.99
14	2.923	5	2	0.63
15	3.326	5	1	0.23
16	3.889	3	1	0.54
17	3.751	2	2	1.31

*Key: Deviation from previous school KCSE chemistry mean grade: 1=negative, 2= Not at all, 3= +1, 4= +2, and 5= +3.

*Key: Frequency of ICT Usage: 5-Daily, 4-Weekly, 3-Monthly, 2-Termly, 1-Not at all

Table 24 presents the summary of individual schools KCSE chemistry mean scores, deviation from the previous school mean as projected by chemistry teachers, and this was compared with the teachers mean frequency of ICT integration. The findings show that teachers in schools number 8 and 17 projected "No deviation" from previous mean, as their KCSE chemistry grades stood at 3.894 and 3.751 respectively. The teachers in both schools posted frequency of ICT usage of 2 meaning the teachers at least integrated ICT termly in pedagogy, and Standard Deviation of 1.72 and 1.31 respectively indicating an almost perfect agreement. Similarly, teachers in schools number 1,2,3,6,10,11,13, and 16 had projected a positive mean deviation of +1 from previous mean for their schools chemistry KCSE mean grade, while from their performance, their mean grades for the years of analysis were 4.207, 3.277, 7.472, 4.489, 3.652, 6.141, 4.262, and 3.889 respectively. All the teachers in these specific schools had a mean frequency of ICT integration ranging from 1 to 3 meaning the teachers did "Not at all" integrate ICT in pedagogy or did it on termly or monthly basis with Standard Deviations ranging from 0.45 to 1.59 respectively. This finding shows that when teachers integrate ICT in pedagogy less frequently they have low expectations.

Table 24 also shows that teachers in schools number 4, 5, 9 and 12 projected a mean deviation of +2 from previous mean, while the chemistry mean grades for the study period were 5.446, 7.472, 5.762 and 4.146 respectively, while their mean frequency of ICT usage ranged from weekly to termly. The findings indicate that when teachers consider integrating ICT more frequently in pedagogy, they develop high expectations for their learners.

On the other hand, teachers in schools number 7, 14 and 15 had projected a mean deviation of +3 from the previous mean before the onset of ICT in their schools whose mean grades stood at 9.383, 2.923 and 3.326 respectively. The teachers in the schools had mean frequency of ICT usage of 4 - weekly, 2 - termly use and 1 - not at all, with standard deviation of less than 1.59 indicating an almost perfect agreement. The findings indicate that some teachers could have projected their expectations without any input in their frequency of ICT usage. The high mean grades in certain schools reflect the frequency that such teacher indicated to integrate ICT in pedagogy. On the other hand the projections on the mean grades did not reflect on the performance in majority of the schools. With this in mind the researcher adopted a sequential exploratory design and revisited the study schools to actually observe teachers integrate ICT in chemistry lessons. Only 29% (five schools out of seventeen) organized hypothetical ICT integrated lessons for the researcher to observe as reported in the previous section; meaning that the lessons were organized but real ICT integration in teaching and learning chemistry does not take place in public secondary schools in Kisumu County. The findings from the lesson observation indicate that ICT integration to teach chemistry in secondary schools in Kisumu County does not take place hence has no effect on performance of such schools.

4.6.2 Chemistry teachers' perceptions of frequency of use of software in teaching and learning on learner performance in the subject

To explore possible implications of the observed chemistry mean grades with regards to teachers' perceptions on ICT integration, an analysis was performed to determine the frequency that teachers of chemistry in the study schools claim to integrate ICT software in teaching and learning. The status of frequency of use was rated as used, then compared against the performance of their students. The summary of teachers' response on the extent of use of specific software in teaching and learning chemistry was presented in Table 25.

Table 25: Chemistry Teachers Perceptions on Frequency of Use of Software in Teaching and Learning chemistry (N=39)

Soft ware	Frequency	of	use			Mean	SD
	5	4	3	2	1		
Internet	19	3	0	1	16	3.21	1.94
E-library	9	0	0	1	29	1.05	1.70
Video shows	8	8	2	10	11	2.79	1.56
Content rich software	12	3	6	4	14	2.87	1.70
Content free software	15	6	2	2	14	3.15	1.80
CD-ROM topical revision	15	3	2	4	15	2.97	1.83
Websites	12	4	2	4	17	2.74	1.79
CD-ROM Encyclopedia	12	0	2	4	21	2.44	1.80
Microsoft Excel	20	5	0	7	7	3.62	1.66
Microsoft office	21	8	0	4	5	3.95	1.49
Learning systems	11	6	2	3	17	2.74	1.77
Training software	12	3	1	2	21	2.56	1.85
Audio-shows	11	0	9	4	15	2.69	1.66
Simulated videos	8	6	8	4	13	2.79	1.56
Animated videos	9	5	7	4	13	2.82	1.61
Topical videos	9	4	6	5	13	2.76	1.62
Computer skills	11	5	9	7	7	3.15	1.48
Interactive learning software	10	4	5	4	16	2.69	1.69
General revision	12	8	3	4	12	3.10	1.68

*Key: 5-Daily use, 4-Weekly use, 3- Monthly use, 2-Termly use, 1-Not at all

Table 25 shows the frequency that chemistry teachers integrate variety of ICT software in their classrooms during teaching and learning. At a glance it was possible to state that the teachers' frequency of use of variety of ICT was generally low because the teachers who responded to integrate ICT on daily and weekly basis were less than 50% in majority of the items presented to them. Examining the individual items for integration, Internet, E-library and websites as a resources from the Internet, were never integrated by 41.03%, 74.35% and 43.58% of the teachers respectively; the means for frequency of use of these software were 3.21, 1.05 and 2.74 respectively, with a standard deviation of less than 2 indicating average use of Internet for integration was on monthly basis or not at all

Video shows, content rich software, content free software, CD-ROM topical revision, general revision and CD-ROM encyclopedia were never integrated by 79.49%, 69.23%, 69.23%, 61.54%, 30.77% and 53.85% of the teachers respectively on daily basis and on weekly basis, the mean for frequency of use of these software were 2.79, 2.87, 3.15, 2.97, 3.10 and 2.44 respectively with standard deviation of less than 2, indicating an almost perfect agreement on average use on monthly basis. General revision software was therefore the most popular revision software among teachers in the study schools.

Microsoft excel and Microsoft Office were the most frequently utilized software for integration by 51.28% and 53.81% of teachers on daily basis; the mean frequency of use of these software were 3.62 and 3.95 respectively indicating an average use on weekly basis. Learning systems and training software enable learners perform peer teaching, but was never at all utilized for integration by 43.59% and 53.85% of teachers in the study schools respectively; the mean frequency of use of these software were 2.74 and 2.56 respectively, with standard deviation of less than 2 indicating a perfect agreement on an average of monthly use.

Audio shows, simulated and animate videos and topical videos revision software   were never utilized for integration by 38.46%, 33.33% and 33.33% of teachers respectively; the mean frequency of integration of these software as claimed by chemistry teachers were 2.69, 2.82 and 2.76 respectively, with standard deviation of less than 2 indicating an agreement on average use of monthly basis.

Software that teaches computer skills was never utilized by for integration by only 17.95% of teachers in the study schools; the mean frequency of use of this software was 3.15 indicating that averagely this software was integrated on monthly basis, with a Standard Deviation of 1.48. It appears to have been the most used software to varied extents by teachers, most probably used to practice computer skills. Interactive learning software was never integrated on daily basis by 41.03% of the teachers in their chemistry lessons; the mean frequency of use of this software was 2.69 indicating an average of monthly use, with a Standard Deviation of 1.48. This finding was deduced to indicate that ICT integration was still minimal, in spite of having revision materials on subject content in storage devices available in the public schools. This finding was deduced to indicate that ICT integration was still minimal, in spite of having revision materials on subject content in storage devices available in Kenyan public schools. Lack of uptake of ICT by Kenyan teachers could be pegged to their lack of exposure to global digital age. Research on the use of data logging for example, in physics and chemistry (Mishra & Koehler, 2011) established the effectiveness of theory and practice of data logging to enhance practical work in chemistry.

4.2.4 Chemistry Teachers' Perceptions on Mean Integration of ICT and Performance

In order to have an index on mean extent of pedagogical integration of ICT by chemistry teachers per school; the teachers' responses on extent of use of ICT in each school was averaged, this was correlated with performance to show pedagogical influence of ICT usage on performance at KCSE. The mean usage of ICT for integration by teachers was computed from their responses on extent of use of ICT, Their responses were then analyzed for teachers in each school and ranked from the least use = 1, to the highest use = 5; then compared with specific school chemistry mean grades. The results of this investigation were presented on Table 26.

Table 26: Relationships between Teachers' Perceived usage of ICT in Pedagogy per school and Chemistry Performance (n=17)

School	School chemistry mean score	Teachers average ICT usage
1	4.207	0.74
2	3.227	2.24
3	7.427	1.16
4	5.446	3.03
5	7.472	1.61
6	4.489	1.18
7	9.383	3.26
8	3.894	1.21
9	5.762	3.09
10	3.652	1.53
11	6.141	2.40
12	4.146	3.25
13	4.262	3.65
14	2.923	1.21
15	3.326	0.5
16	3.889	2.00
17	3.751	0.53
Average	4.845	2.00

*Key: 5-Daily use, 4-Weekly use, 3-Monthly use, 2-Termly use, 1-Not at all

Table 26 gives a summary of the relationship between teachers mean proficiency of pedagogical integration of ICT and performance in chemistry per school. The teachers' mean proficiency was calculated from the average of chemistry teachers' proficiency and use of ICT per school. The schools' mean score in chemistry was calculated from performance per school as obtained from the observation checklist. The Table shows that teachers whose mean pedagogical proficiency integration of ICT is ranked at below 0.5, rated as least use of ICT came from school number 15 whose mean score was 3.326, while teachers whose mean pedagogical integration of ICT was between 0.53 to 1.21 rated as used ICT on termly basis came from schools number 17,1,3,6 and 14 whose scores were 3.751, 4.207, 7.472, 4.489 and 3.894 respectively. The frequency of use of ICT and performance does not reflect any relationship.

Teachers whose mean usage of pedagogical integration of ICT was between 1.53 and 2.40 rated as used ICT termly came from schools number 10, 5, 16, 2 and 11 whose mean grades were 3.652, 7.472, 3.889, 3.277 and 6.141 respectively. Teachers whose mean integration of ICT was between

3.03 and 3.26 rated as used ICT monthly came from schools number 4, 9, 12 and 7 whose mean scores were 5.446 5.762, 4.146, 9.383 respectively. Teachers whose mean integration was 3.65 rated as used ICT Daily came from a school whose mean score was 4.262. The average ICT proficiency for the teachers in the study was 2.00 which indicated that averagely, the teachers in the study tended to integrate ICT in pedagogy on termly basis.

The results of analysis in this sub-section indicate that performance could not be pegged on frequency of use of ICT by teachers concerned, as other factors like the type, culture, environment and other infrastructural matters of a school also impact on performance to varied extents. For this reason the schools were not ranked based on performance as type of school dominated as an extraneous variable affecting performance. This finding is comparable to a study by Adel and Mounir (2013) that focused on direct and indirect effects of ICT on standard explanatory factors, stating that student performance is explained by students' characteristics, educational environments and teacher characteristics. Stating further that ICT may impact those determinants and consequently the outcome of education. The differences observed on student performance are thus more related to differentiated impact of ICT on standard explanatory factors. This contrasts with the current study focusing mainly on teacher factors, specifically teachers' perspectives in integration of ICT on learner performance. As can be seen in Table 25, the mean frequency of use of ICT by teachers in a school is not necessarily reflected in the mean performance of their students in chemistry, but there is a possibility of influence of several factors.

This finding is closely in tandem with the findings from the study by Qing (2010) who investigated In-service chemistry teachers' attitudes towards ICT in Xi'an, considering computer competence, computer attributes, culture, perception and personal characteristics as independent variables, he established that the degree of participants' computer attitudes is in between neutral and positive, and was significantly related to computer attributes, competence, culture, perception and performance. Asserting teachers play an important role in information, society and placing high emphasis on communication skills. Adding that, ICT integration at school will have little impact if teachers are not actively involved in their integration to the curriculum.

There is therefore no doubt that teachers are at the center of success if any, of the integration of ICT. The present researcher considers a teacher in the light of Mishra *et al.* (2006) in their framework for teacher knowledge and learning as imparter of knowledge, designer and counselor, coordinator, participant, explicator and assessor. As facilitators in learning, teachers' perceptions may affect the use of ICT directly and affect students' performance in science indirectly. It is therefore important that teachers have positive attitude towards ICT proficiency and usage. The quality assurance officers' perceptions were sought. The SCQASO were therefore asked to give their perspectives on the extent ICT integration has influenced syllabus coverage, KCSE mean grade and KCSE chemistry mean grade in ESP schools.

A summary was made on the opinion of SCQASO on the influence of ICT on syllabus coverage; KCSE chemistry and school KCSE mean score. One expert was of the opinion that ICT had influenced syllabus coverage to a great extent as learners interacted with computers to extend learning sessions. He also opined that improvement had been noted on general mean score and chemistry mean grade at KCSE in many schools. Another expert was of the opinion that ICT had influenced KCSE mean grade to some extent. One expert gave an opinion of average influence on all aspects of ICT. While another expert was of the opinion that ICT had made syllabus coverage easier and faster and the mean grades in KCSE and in chemistry had improved positively.

On the other hand another expert was of the opinion that there was low embracement of ICT on syllabus coverage, while the influence on KCSE mean grade was remarkable; but was of the opinion that influence on chemistry mean grade was not yet as remarkable. Another expert on syllabus coverage was of the opinion that ICT had enabled learners to cover more content and interact with new ideas online, while the influence on KCSE mean grade had shown improvement; but was of the opinion that there was still a problem with chemistry in most schools, without clarifying what type of problem was being experienced by the schools.

These mixed reactions by the experts indicate that ICT integration influence was not yet universal in the county as each sub-county expert had own independent opinion on the influence of ICT on the aspects of syllabus coverage, KCSE and chemistry mean grade. Perhaps this is an indication of lack of networking among the county personnel on the process of ICT integration, introduction of departmental seminars and workshops could serve to reduce the differences.

This finding adds to literature surrounding perceptions of teachers on traditional versus technology enhanced learning environment, suggesting that the stakeholders view technology as effective but do not engage in its effective uptake. This finding is in agreement with the views of Lindfores

(2007) who investigated perceptions of European teachers on the use of ICT in teaching; and established that teachers have a perception that ICT use has to be combined with traditional classroom based teaching from pedagogical point of view, and the impact of ICT in education greatly depends on how it is used. This finding is in agreement with the views of Sagra and Gonzales (2017) who acknowledge the role of ICT in promoting learner attention and improvement of perception skills; but observed that teachers have not yet discovered or understood the possibilities that ICT offers to students as a means of complementing their traditional receiver role with that of message producer-transmitter, a view that is also held by Cuban *et al.* (2001) Drent and Melisen (2008). The current studies agrees with the view that teachers have not embraced ICT integration, but only regard it as effective on improving learner performance from their perspectives, but not from their practice.

In summary performance trends in the study schools did not reflect on ICT integration shown by perceptions of teachers on frequency of use of ICT in teaching and learning chemistry. Teachers who indicated to integrate ICT on daily and weekly basis are less than 50% in almost all aspects of integration in pedagogy that was presented to them. Microsoft word and Microsoft excel are the most frequently used software by over 50% of teachers than any other software available; this can be attributed to the fact that teachers use these software for administrative duties like compiling marks and creating worksheets instead of pedagogy.

Teachers' frequency of use of ICT software did not reflect their perceptions on performance, as some of those teachers who indicated to be least users on termly basis or not at all, reflected high performance on their learners; while those teachers who indicated as most frequent users on daily or weekly basis, the performance of their learners did not measure to equal standards. Therefore integration of ICT in teaching and learning chemistry is not practiced in secondary schools in Kisumu County; hence it has no effect on performance. Ministry of education, through quality assurance and standards officers, SCQASO perceptions on the subject was that ICT integration in teaching specific subjects like chemistry in Kenya was not yet a universal practice, and as a resource for teaching and learning there is expected to be improvement on performance.

CHAPTER FIVE

SUMMARY, CONCLUSIONS AND RECOMMENDATIONS

5.1 Overview

In this chapter, the summary of major findings, conclusions and recommendations of the study on Information Communication Technology preparedness, Integration in Education and Stakeholders perspectives on Chemistry Performance among Public Secondary Schools in Kisumu County, Kenya have been presented. The presentation has been done according to the study objectives.

5.2 Summary of Findings

The study produced the following findings:

5.2.1 Availability of ICT resources for teaching Chemistry in Public Secondary Schools in Kisumu County

The first objective of this study was to assess availability of ICT resources for teaching chemistry in public secondary schools in Kisumu County. Data analysis and interpretation of questionnaire responses from the school principals, chemistry teachers, chemistry students and SCQASO and observation schedule established that the quantity of computers and other ICT equipment, resources and accessories including hardware and software associated with ICT integration were inadequate and when available, were only found at the ICT centers or computer laboratories; hence not sufficient for integration in teaching chemistry at the study schools.

5.2.2 Chemistry Teachers Capacity to Integrate ICT in Pedagogy

The second objective of the study was to establish chemistry teachers' capacity to integrate ICT in pedagogy. Data analysis and interpretation of questionnaire responses from school principals,' teachers and SCQASO, observation schedule and lesson observation checklist revealed that all teachers in the study schools are computer literate, and have acquired skills on use of technological tools, but none indicated in their lesson plan or schemes of work that they integrate ICT in teaching the subject, hence skills to integrate technology in pedagogy haven't been sufficiently acquired; over 72% of teachers frequently performed administrative tasks with ICT, but seldom performed any tasks in lesson preparation or pedagogy with ICT. Lesson observation revealed that teachers do not practice ICT integration in pedagogy.

5.2.3 Extent of Integration of ICT Resources in Teaching and Learning Chemistry in Public Secondary Schools in Kisumu County

The third objective of this study was to assess the extent of integration of ICT in teaching and learning chemistry in public secondary schools in Kisumu County. Data analysis and interpretation of questionnaire responses from chemistry teachers, chemistry students, SCQASO, observation checklist and lesson observation checklist revealed that ICT enhanced environment does not exist in Kenyan classrooms as only 24.5% of learners in the study schools had experienced some form of ICT integration in their chemistry lessons; and over 75.5% of learners had never experienced teaching and learning of any chemistry topics integrated with ICT.

Lesson observation of ICT integrated chemistry lessons at the study schools failed to take place in 70% of the schools, hypothetical ICT integrated classes were organized for observation in 30% of the schools, indicating that routine ICT integration is not practiced. These findings show that ICT integration has not been adopted or accepted by teachers in public schools.

5.2.4 Stakeholders Perceptions on the Influence of ICT Integration on KCSE Chemistry Performance

The fourth objective of this study was to determine stakeholders' perceptions on ICT integration with regards to KCSE chemistry performance. Data analysis and interpretation of questionnaire responses from principals, teachers and SCQASO revealed that teachers seldom used ICT in pedagogy; but occasionally used Microsoft word and Microsoft excel to compile marks and create work sheets, tasks which are not pedagogical. These findings indicate that ICT integration in teaching and learning is based on teachers' perceptions, and will only be successful if teachers put into practice their positive perceptions on the outcomes of technological use on learner performance.

5.3 Conclusions

This study investigated ICT preparedness, integration in education and stake holders' perspectives on chemistry performance among public secondary schools in Kisumu County, Kenya. This was in relation to the need to devise methods of teaching chemistry integrated with ICT to make its abstract concepts easier to the level of understanding of secondary school students to improve their performance. The presentation has been done according to the study objectives. Based on the findings of the study, the following conclusions have been arrived at:

5.3.1 Availability of ICT resources for teaching Chemistry in Public Secondary Schools in Kisumu County

In its first objective the study specifically sought to assess the availability of ICT resources in teaching chemistry in public secondary schools in Kisumu County. The study established that ICT resources available at the computer laboratories or ICT centers in the study schools are not adequate to be used for integration in teaching chemistry. In view of this finding, the study concludes that traditional educational learning methods still dominates Kenyan education system, where the pupils' principal learning resource is the teacher, and role of technology is revision of subject matter; therefore ICT does not influence performance of the subject.

5.3.2 Chemistry Teachers' capacity to Integrate ICT in Pedagogy

In its second objective, the study sought to establish chemistry teachers' capacity to integrate ICT in pedagogy. The study established that all chemistry teachers in the study schools are computer literate but hardly perform any tasks in lesson preparation or in pedagogy with ICT. In view of this finding the study concludes that it is the traditional method of teaching chemistry that influences performance in the subject and not teachers' ICT literacy. This means that no matter how high the level of competence and confidence of teachers, ICT will never be integrated into normal classroom practice until the technology is available and accessible in the classrooms where teachers teach.

5.3.3 Extent of Integration of ICT in Teaching and Learning Chemistry in Public Secondary Schools in Kisumu County

In its third objective, the study specifically sought to assess the level of integration of ICT resources and learner performance in chemistry among public secondary schools in Kisumu County. The study established that ICT integration in teaching and learning chemistry is not practiced to any significant extent in the study schools in Kisumu County to have any impact on performance of the subject. In view of this finding, the study concludes that a paradigm shift is required to encourage Chemistry teachers to adopt modern teaching methods in curriculum delivery.

5.3.4 Stakeholders' Perceptions on Influence of ICT Integration on KCSE Chemistry Performance

In its fourth objective the study sought to establish stakeholders' perceptions on influence of ICT integration on KCSE chemistry performance. The study established that even though stakeholders have positive perceptions on ICT integration towards performance of learners at KCSE examinations, teachers seldom perform ICT tasks in pedagogy. In view of this finding the study concludes that perceptions of stakeholders on ICT integration are opinions not put into practice, hence do not impact on performance of learners in chemistry.

5.4 Recommendations from the Study

The researcher has argued in this report that ICT resources at ICT centers or computer laboratories, levels of integration of ICT in pedagogy, chemistry teachers' capacity to integrate ICT in pedagogy and the stakeholders' perceptions on the influence of ICT integration have no influence on performance of the subject in the study schools. The study has also shown that ICT resources are not being utilized for integration in teaching and learning chemistry; and that despite chemistry teachers being IT literate and have positive perceptions to ICT integration, ICT as a resource for teaching and learning has no impact on KCSE chemistry performance among public secondary schools in Kisumu County. It is against this background that the recommendations below are made. Basing generalizations on the findings of this study, the researcher recommends the following based on the study objectives that could help improve the state of ICT integration in education in secondary schools in Kenya from facts noted during the study.

5.4.1 Availability of ICT Resources for Teaching Chemistry in Public Secondary Schools

The government of Kenya, through the ministry of education should give priority to construction and improvement of ICT infrastructure and resources in each public school; with the same emphasis that is given to construction and improvement of classrooms, libraries and science laboratories using the ministry Free Day Secondary Education (FDSE) funds granted to public schools. This should enable each school in the republic to acquire enough computers at the ICT center and at least one computer in each classroom and science laboratory, accompanied with a projector and a smart board for presentations to wider audience. This will enable teaching with computers take root in public schools in Kenya. Moreover, when teaching with computers take effect, learners should be encouraged to be innovative enough to search for information on-line using any available ICT resource which should include their mobile phones, tablets, radio and television sets at home; to ease e-learning especially when schools close down due to a global pandemic as witnessed during Covid-19 lockdown.

5.4.3 Teachers Capacity to Integrate ICT in Pedagogy

Teachers should seek training not only on technology use, but also use of technology to teach school subjects, and learn from one another how to gather materials from the web and customize it to use in their lessons. The teachers should cast concepts on the screens to enable students learn better with numerous presentations and pictures. This will make work easier for teachers and enjoyable for the students. Above all, teachers should post their lesson notes, students' assignments and progress reports in the school website. Through these processes teachers and students will get used to doing school work with computers.

5.4.4 Extent of Integration of ICT in Teaching and Learning Chemistry in Public Secondary Schools

Every school in the republic must have a website and portals for learners to access learning materials from their smart phones, laptops, tablets or personal computers. In addition, every school should develop a school ICT policy based on the national ICT policy in education. The school policy should include among other issues, that all teachers and any new teacher posted to the school must attend evening computer literacy classes for at least two months to acquire basic computer skills, then train his or her students on computer use in his subject. The students in turn teach one another after classes how to work with computers in the computer lab with teacher on duty supervising.

5.4.5 Stakeholders Perceptions on Influence of ICT integration on Chemistry Performance

Stakeholders in the ministry of education and private sector should regard critically the importance of the integration of ICT in the Kenyan curriculum. This couldn't have been clearer with the onset of lockdown after Covid-19 pandemic that paralyzed any and all learning activities across all planes and divide in the education sector. With no way of accessing learning and teaching materials, all forms of learning came to a halt, including universities. The situation could have been salvaged with skills from ICT integration. Teachers can also post and evaluate assignments also posted on-line, all without the risk of exposure to Covid-19. ICT is the only plane by which the virus can't spread, that is why many resorted to working from home on-line.

Stakeholders at the ministry headquarters and at county levels should find ways to assist learners acquire laptops or tablets or personal computers for school work at home at all times including school holidays

5.5 Suggestions for Future Research

Having gone through the study on influence of teacher preparedness in integration of ICT on performance in chemistry among public secondary schools in Kisumu County, Kenya, the researcher realized that there is need to do further investigation on the following areas:

1. More research should be done on on-line teaching and learning activities of all school subjects such that learning in Kenyan schools can successfully be accessed on line in specific school websites even during lockdown as experienced during Covid-19 pandemic that affected the whole world.

2. More research should be carried out to identify ways of achieving successful implementation of ICT integration in teaching and learning process in Kenyan public schools by identifying specific impediments towards its success, especially mode of training teachers on technological knowledge; and how to overcome the impediments.

3. A survey should be carried out to identify and document the development of modern pedagogical orientations and key ICT players in the Kenyan education system, so that policy makers may have factual information on equipment type and basic needs, and hence form a basis for decision making in allowing more ICT players in the education field.

4. A survey should be carried out to determine how Kenyan schools' activities can be made digital on the schools' platforms in form of school websites, such that all school teaching and learning notes, reporting, evaluation and information are accessed on-line just like in the new ministry of education information system referred to as NEMIS.

REFERENCES

Abbit, J. T. (2011). Measuring Technological Pedagogical Content Knowledge on Pre-Service Teacher Education: A Review of Current Methods and Instruments. *Journal of Research on Technology in Education 43* (4)

AbdelKarim, O. Ahmed, L. Hassane, D. & Khalid L. (2012). *ICT Integration in Chemistry-Physics classes in Middle Schools through a Participatory Pilot Project Approach* School of Science and Engineering: Al Akhawayn University.

Ifrane 53000 Morocco Abdelrahman, A., Howie, S. & Osman (2013). *The Integration of ICT in Teaching Science and Mathematics in secondary school; with particular Reference to Sudan* http://www.ub.es/multimedia/iem

Afashari, M. B. & Samah, W. (2009). Factors Affecting Teachers' Use of ICT *International Journal of Instruction 2* (1), 77-104.

Alev, N. (2003). *Integrating Information Communication Technology (ICT) into Pre-Service Science teacher Education: The Challenges of Change in a Turkish Faculty of Education* https://www.sematicscholar.org>1

Alvardo, G. R. (2011). *Caraga Regional Science High School: On the Rise*. Available on line at https://sites.google.com.karagdagang.pangyari.

Alazam, A. O., Baker, A. R., Asmiran, S. (2012). Teachers' ICT Skills and ICT Integration in the Classroom: The case of VocTIONAL AND Technical Teachers in Malaysia. *European Journal of Special Education Research 2501*

Anderson, J. A. & Eloumin, K. (2004). *Philosophies and Philosophical Issues in Communication.* https://onlinelibrary.wiley.com>labs.

Arinto, P. (2005) *National Strategic Plan for ICTs in Basic Education Initiatives* http://ejournal.eduprojects.net

Ashley, W. (2016). *Ten Reasons Today's Students need Technology in the Classroom.* http//www.securedgementworks.com//10reasonstoday's-students-need-T

Atherton L.A. (2011). *Factors Affecting Students Performance in Chemistry* https://researchgate.net/profile/Ahmed.

Avinash, A. & Shailja, S. (2013). The Impact of ICT on Achievement of Students in Chemistry at Secondary level of CBSE and up board in India *International Journal of Science and Research 2* (8) 126-129

Ayere, M.A. (2009). *Comparison of Information and Technology Application in NEPAD and non-NEPAD schools in Kenya* PhD Thesis: Maseno University.

Badeleh, A. & Sheela, G. (2016). Effects of ICT based Approach and Laboratory Training Model of Teaching on Achievement and Retention of Chemistry *Journal of Contemporary Educational Technology 2* (3) 213-237.

Balcon, P. (2003). *Information and Communication Technologies (ICT) in Teacher Education (ITE) Programs in the World and in Turkey (a Comparative Review)* https://www.sciencedirect.com>pi

Baker, S. (2018). *Best Practices for Data Analysis of Confidential Data.*https;//ria.princeton. edu.best-practice. Princeton University 87 Prospect Avenue Barak, M. (2008). *Transition from Traditional to ICT-enhanced learning Environments in Undergraduate Chemistry Courses* www.sciencedirect.com.

Barak, N. (2007). Transition from Traditional to ICT-enhanced Environment in Undergraduate Chemistry Courses. *Journal of Science Education 89* (1) 117-139.

Barnea, N., Dori, Y. J. & Hofstein, A. (2010). Development and Implementation of Inquiry-Based and Computerized-Based laboratories: Reforming High School Chemistry in Israel *Journal of Chemistry Educational Research and Practice 11* 218-228

Batchelor, S., Evangelista, S., Hearn, S., M. Sugdem, S. & Webb, M. (2003). *ICT for Development: Contribution to the Millennium Development Goals* Washington DC

Basri, W., Alandejani, A. J., Almadani, F.M. (2018). ICT Adoption impact on Students Academic Performance: Evidence from Saudi Universities. *Journal of Hindawi Education Research International 2018* Article ID 124 0197

Beaurain, C. (2016). *Life Energy Consultant* https:www.theshieldoflife.com

Becta, T S. (2005). *Evidence on the progress of ICT in Education.*

Becta.ICT Research. www.becta.org.uk/research/html

Bell, M. (2008). Ethics in Child research: Rights, Reasons and Responsibilities. *Journal of Children Geographies 6* (1) 7-20

Bell, T., Urhahne, D., Schanze, S. & Ploetzner, R. (2010). Collaborative Inquiry Learning: Models, Tools and Challenges. *International Journal of Science Education 32* 349-377.

Bernea, N., Dori, Y. J. & Hofstein, A. (2010). Development and Implementation of Inquiry-based Laboratories: Reforming High School Chemistry in Israel. *Journal of Chemistry Educational Research and Practice 11* 218-228

Bethesda, M.D. (1978). Report: Ethical Principles and Guidelines for the Protection of Human Subjects of Research. The Commission-University of Kansas Medical Center.

Berg, B.L. (2004). Qualitative Research Methods for Social Sciences. 5th Edition, Pearson Education, Boston. *American Journal of Industrial and Bussiness Management 5* (5). https://www.scirp.org>references

Betz, M. (2011). ICT in School: The Principal's role. *Journal of Educational Technology and Society 3* (4)

Bevernage, A., Cornille, B. & Mwaniki, N. (2005). *Integrating ICT in Teacher Training, Reflections on Practice and Policy Implications- A Case Study of the Learning Resource Center the Kenya Technical Teachers' College* htpp://www.idrc.ca/en/ev93060-201-1-DoTo.p

Bhattacharya, I. & Sharma, K. (2007). India in the Knowledge Economy- an Electronic Paradigm *International Journal of Educational Management 2* (6), 543-568.

Bitner, N. & Bitner, J. (2002). Integrating Technology in the Classroom; Eight Keys to Success. *Journal of Technology and Teacher Development Literacy Information and Computer Journal* (LICEJ) *6* (2)

Bloor, M. & Wood, F. (2006). *Key Words in Qualitative Methods: A Vocabulary of Research concepts* London: Sage publishers

Bogdan, R. C. & Biklen, S. K. (1992). *Qualitative Research for Education: An Institution to Theory and Methods* Ally and Bacon

Bonfilius, V. (2019). *Integration of ICT in Teaching-Learning* https://www.slideshare.net>mobile>inte.

Burewicz, A. and Miranowicz, N. (2006). Effectiveness of Multimedia laboratoryInstruction Department of chemistry Education, Faculty of Chemistry University of Poland *Journal of Chemical Education Research and Practice* 2006, 7 (1).

Burton, J. K., Moore, D.M. & Magliard, S. (2004). *Behaviorism and Instructional Technology. Handbook of Research on Educational Communication and Technology.* London: New Jersey, Lawrence Erlbaum Associates, Publishers, 3-3.

Butzin, S. (1988). *Project CHILD: A Decade of success for young Children* https://www.theejournl.com/magazine/vault/A2882.cfm

Cabanatan, P. G. (2001). *ICT Trends in Teacher Training Curricular: An Asia-Pacific Perspective* Paper presented during the Asia-Pacific workshop on Educational Technology in Tokyo Japan. http://gaugeougakugei.ac.jp/apei01/papers/cabanatan.htm

Cabanatan, I. (2001). Towards a Framework for the Use of ICT in Teacher Training in Africa. *Journal of Distance and E-learning 20* (2) https://www.tandfonline.com>do

Camacho, V.J., Camacho, R.R., & Basilisa, V. (2012). *Some Imperatives of ICT Integration in the Philippine Education System: Towards Modernization and Relevance in Highly Globalised Economy* http://education.gov.ats.ca/k 12/curriculum by subject

Carlson, S. & Gadio, C.T. (2002). *Teacher Professional Development in the use of Technology: Potentials, Parameters and prospects.* Washington D.C. http://unescodoc.unesco.org/images/0012/001234.2e.pdf.

Carpi, A. & Mikhailova, Y. (2014). *The Vision learning Project: Evaluating the Design and Effectiveness of Interdisciplinary Science Web content* J. College science online htt://www. visionlearning.Com/library/teach.

C.D.E (2017). County Director of Education, Kisumu County Secondary Schools Analysis

CEMASTEA, (2018). Secondary INSET manual: *Training Manual for National INSET- 2018.* Nairobi: CEMASTEA.

CEMASTEA, (2012). Secondary INSET manual: *Training Manual for National INSET- 2012.* Nairobi: CEMASTEA.

Cadellini, L. (2012). *Chemistry: Why the Subject is Difficult.* https://doi.org/10.1016/s0187-893x(17) 30158-1

Chai, C. H., Ho, J. N H., & Tsai, C.C. (2012). Examining Pre-Service Teachers'Percieved Knowledge of TPECK and Cyber-wellness through Structural equal model*Australian Journal of Educational Technology. 28* (6).

Chai, C.S., Hong, H. Y. & Teo, T. (2009). Singaporean and Taiwanese pre-service Teachers beliefs and their attitudes towards ICT: A Comparative Study. *Journal of the Asia-Pacific Education Research. 18* 117-128.

Champman, O., Chich, H.L. & Vincent, J. L. (2012). Constructing Pedagogical Knowledge of Problem Solving: Pre-service Mathematics Teachers. University of Calgary. *Journal for Research in Mathematics Education 5* (21) 132-144

Cohen, D. & Hill, H. (2001). *Learning Policy: When State Education Reform Works* Yale University Press. http://www.researchgate.net>2712

Cohen, L. & Manion, L. (2006). *Research Methods in Education* Croom-Helm Limited. https://www. Researchgate.net>2718

Colin, H. Lunzer, E.A., Tymms, P., Tylor, F.C., & Restorick, J. (2004). *Use of ICT and its relationship with Performance in Examinations: A Comparison of the Impact CT2 Project's Research findings using Pupil-level, School-level and Multilevel modeling Data academia.edu*/12496064/use-of-ict-and-its-impact.

Cornille, B . & Mwaniki, N. (2002). Teachers' Gender Influence on Adoption and Use of ICT in Public Secondary Schools in Kenya *African Journal of Education Science and Technology ajest.info. >ajest>article>view*

Creemers, S. & Kyriakides, L. (2008). The Effects of Teacher Factors on Different Outcomes: Two Studies Testing the Validity of the Dynamic Model. *Journal of Effective Education1* (1) 61-85. https://www.tndfoneline.com>abs.

Creswell, J. W. (2003). *Research Design: Qualitative, Quantitative and Mixed Methods Approaches* (2nd Edition) Thousand Oaks, C A: Sage

Cuban, L. Kirkpatrick, H., Peck, C. (2001). High Access and Low Use of Technologies in High School Classrooms: Explaining an Apparent Paradox. *Amertrican Educational* Journal 2001 *38* (4), 813-834.

Davis, N.E. & Tearle, P. (1999). *Acore Curriculum for Telematics in Teacher Training*

Decristan, J., Klieme, E., Kunter, M., Hochweber, J., Buttner, G., Fauth, B. & Hardy, I. (2015). Embeded Formative Assessment and Classroom Process Quality: How do they Interact in Promoting Science Understanding? *American Educational Research Journal 52* 1133-1159

Davis, N., Preston, C. & Sahin (2009). Teacher Training: Evidence for Multinivel Evaluation from a National Initiative *British Journal of Education 40* (1), 135-148.

Demiray, U. (1990). Diffusion of Distance Education in Turkish higher Education *Journal of Educational Technology Research 45* (122) 124-128. https://link.springer.com>article

Denisia, S.P. & Suresh, J. K. (2013). *Factors influencing the Effective use of Computer in Teaching Chemistry* https:www.languageinindia.com

Dori, Y. J., Rodrigues, S. & Schanze, S. (2013). *How to Promote Chemistry Learning through the Use of ICT.* Teaching Chemistry-A Study book Sense-Publishers, Rotterdam. https://doi.org/10.1007/978-94-6209-140-5-8

Draper, K. (2010). *Understanding Science Teachers use and Integration of ICT in a Developing country Context* University of Pretoria https://repository.up.ac.Za>t

Drent, M. & Meelissen, M. (2008). Which factors Obstruct or Stimulate Teacher Educators to use ICT innovatively? *Journal of Computers and Education* (2008).*51* 187-199.

Earle, R. (2002). The Integration of Instructional Technology into Public Education: Promises and challenges *Journal of Educational Technology42* 5-13.

Enginda, T. (2012). *ICT enhanced Teacher STDs for Africa.* Ababa: UNESCO-11CBA. http://www.eng.unesco-iicba.org/sites/default/files/pdf.

Ertmer, P. (2005). Teacher Pedagogical Beliefs: The Final Frontier in our Quest for Technology Integration *Journal for Educational Development, Research and Development 53* 25-39.

Eurydice, K. (2004). *Key Data on ICT in Schools in Europe, 2004 edition.* http://www.eurydice.org/resources/pdf/o-int

Eyup, T. & Volkan, D. (2017). Teachers' attitude Towards Liquid Crystal Display (LCD) Panel Interactive Board Applications. *Academic Journals 12* (23) www.academiajournal.org.ERR

Fauth, B., Decristan, J., Rieser, S. & Buttner, G. (2014). Student ratings of Teaching Quality in Primary Schools: Dimensions and Predictions of Students Outcomes. *Journal of Learning and Instruction 1* (29) 1-9. https://www.researchgate.net>2591

Ferrell, G., Isaac, C., & Trucano, M. (2007). *The NEPAD e-schools Demonstration Project: A Work in Progress. A Public Report* Washington: Info.dev.

Fouts, J. (2000). *Research in computers and Education: Past, Present and future.* Available at http://www.gatesfoundation.org/nr/download/ed/evolution/computer- researh-Summer.pdf

Fullan, M. G. (1992). Teacher Development: Teacher Beliefs, diversified Approaches and Processes. *Journal for Educational Leadership 50* (6) https://wwwmichaelfullan.ca>2016pdf

Gall, M. D., Borg, W.R. & Gall, J. P. (2007). *Educational Research-An Introduction (9 th Ed.).* Longman.Gilbert, P. (2015). *Tshwane leads Metros in Wi-Fi access.* http://www.itweb.co.za/index.php.option=com%5F

Gay, L. R. (1996). *Educational Research: Competencies for Analyses and Application (5th Ed.)* New Jersey. Prentice Hall

Gilbert, B. (2015). "Online Learning: Revealing the Benefits and Challenges". *Journal of Education Masters 303* https://fisherpub.sjfc.edu>education

Grabe, M & Grabe, C. (2007). *Integrating Technology for Meaningful learning* 5th Edition NY Haughton Miffin

Graham, R., Burgoyne, N., Cantrell, P., Smith, L.; St. Clair, L. & Haris, R. (2009). Measuring the TPACK Confidence of In-service Teachers *Journal of Technological Trends 53* (5) 70-79.

Grant, G. (2002). *Regional Initiative for Information Strategies: Sectoral ICT Strategies Planning Templates* Ottawa: Common Network of IT for Development and the Commonwealth Secretariat.

Government of the Republic of Kenya (GOK) (2007). *Kenya Vision 2030* Government Press, KenyaGood, T. L., Wiley, C.R. & Florez, I. R. (2009). *Effective Teaching: An Emerging Synthesis.* International Handbook of Research on Teachers and Teaching P 803-816. New York: Springer.

Guzey, S.S. & Roehrig, G. H. (2009). Teaching Science with Technology: Case Studies of Science Teachers Development Technology, Pedagogy and Content Knowledge. *Journal of Contemporary Issues in Technology and Teacher Education 9* (1)

Haddad, W.D. & Drexler, A. (2002). *Technologies for Education: Potentials Parameters and Prospects* www.UNESCO.org>resiurces>full-list

Hattie, J. (2009). *Visible Learning: A Synthesis of Over 800 Meta-Analyses relating to Achievement.* USA: Routledge.

Hawkins, R. J. (2002). *The Global Information Technology Report* 2001-2002. New York-Oxford: Oxford University Press.

Hawkridge, G. (2002). *New Information Technology in Education* Sydney: Groom helm.

Haris, S. (2002). Innovative Pedagogical Practices using ICT in Schools in England. *Journal of Computer Assisted Learning* 18(4):449-458. http://www.cris.com/~faben/phil-1.

Hennesey, S. & Onguko, B. (2009). *Developing use of ICT to Enhance Teaching and Learning in East African Schools: A Review of Literature.* A paper presented at Zanzibar Round Table http://www.schoolnetafrica.org

Higgins, S. (2010). *Does ICT Improve Teaching and Learning in Schools?* A British Educatinal Research Assisstant. http://www.bera.ac.uk/files/reviews/ict-pur-mb-r-f- p-laugg03pdf.

Higgin, S. & Moseley, D. (2011). Teacher Thinking about ICT and Learning: Beliefs and Outcomes. *Journal of Teacher Development 5* (2) 191-210.

Hussain, I., Suleman, Q., Nasser, U. D. & Farhan, S. (2017). Effects of ICT on Students Academic Achievement and Retention at Secondary Level *Journal of Education and Educational Development* *https*://files.eric.ed.gov>fulltext

International Society for Technology in Education (ISTE)(2006). Information Technology: Underused in Teacher Education. https://www.mff.orgedtech.

InfoDev. (2020). *Impact of ICT on Learning and Achievement:A Knowledge Mapon ICT in Education* http;//www.infodev.org.>article

Ivankova C. Y. (2011) *Applying Mixed Methods in Action Research*-Sage publications. Available on-line at www.sagebu.com/sites

Ivankova, N. V., Vicky, L., Clerk, P. (2005). *Mixed Methods Applications in Action Research* Available online at us.sagepub.com/en-us/nam/author/nataliya

Isaboke, G. H. (2014). Teacher Preparedness in Integrating ICT in Lower Primary Schools in Nyamira County, Kenya Kenyatta University https://www.djresearch.co.uk>item.

ISTE. (2004-2007). *Future In-service Teacher Training across Europe* http://iste.ssai.valahia.ro

ISTE (2000). *Technology Performance Profiles for Teacher Preparation, International Society of Technology in Education* http://cnets.iste. Org/index3.html

Jack, S. (2016). *Tshwane Priorities: Internet for Development.* http://imnews.co.za/tshwane-priorities-internet-for-development.

Jager, A. A. & Lockman, A. H. (1996). *Impacts of ICT in Education: The role of the Teacher and Teacher Training* University of Leeds www,leeds.ac.uk/educol/documents/0000201.htm.

Jatilen, M. & Cloneria, N.J. (2018). *Teachers Perception on the Use of ICT in Teaching and Learning: A Case Study of Namibian Primary Education.* https://epublications.uef.fi>u.

Java, T. (2004). *Technology, Attitudes, Competencies and Use in Practice Teaching: Implications to Pre-service Teacher Education* Master's Thesis UP-Diliman

Jekins, J., Lieberg, S. & Stieng I. L. (1998). *Socrates-mailbox: the Connected Teacher: Using ICT in School for Teacher Training.* Oslo, National Center for Educational Resources

Jonassen, D. H. (2004). *Handbook of Research on Educational Communication and Technology.* London: New Jersey, Lawrence Erlbaum Associates, Publishers.

Jones, A. (2003) *ICT Competencies Checklist.* https://ictpd.net.

Juuti, K., Lavonen, J., Aksela, M. & Meisalo, V. (2009). Adoption of ICT in Science Education: A case study of communication channels in Teachers' Professional Development project *Eurasia Journal of Mathematics, Science and Technology 5* (4) 103-118.

Karenji, O. J. (2016). Integration of Information Communication Technology in Teaching of English in Secondary Schools in Nyakach sub-county, Kisumu county http://ir-library.ku.ac.ke/handle/123456789/15343..

Katitia, M. D. (2012). *The Role of ICT Integration into Classroom in Kenya.* Available on-line at www.academia education.go.ke

Karsenti, T., Collins,S., Harpen, M. T., (2012). *Pedagogical Integration of ICT: Success and Challenges from 100+ African Schools* ON: IDRC. Http//www-icd/open.ac.uk.ins/results

KCSE (2018). *Reports.* www.kcse-online.info/physicsreprt.html-www.kcse-onlineinfo/biologyreport.html
-www.kcse-onlineinfo/chemistryreport.html

Kembouri N. P., Lausnne P. F., Nemaca, P. K (2004). *Effective Teaching and Leaning using ICT.* Available on line at eprints.iou.ac.uk/705/1/mellvnkennedy, T. & Cox, T. (2008). *Students reading achievement during the transition from primary to secondary.* https://books.google.co.ke

Kent, N. & Facer, K. (2004). Different Worlds? A comparison of Young Peoples Home and school ICT use *Journal of Computer Assisted Learning 20* 440-455.

Kerlinger, F. N. (2003). *Foundation of Behavioural Research* Oxford: Holt Richard and Winstroke

Kidombo, M., Gakuo T. & Kindachu. G.(2011). *Closing the Chasm: Are secondary School Teachers in Kenya Using ICT Effectively to Deliver Curriculum Content?* University of Nairobi, School of Continuing and Distance Education

K.I.E. (2009). Primary Teacher Education: Syllabus book Vol 1. New Ed. Nairobi: Government Printers

K.I.E. (2007). *ICT Integrated Lessons.* http://dimdima.com science common/show.science.asq

Kilieme, E., Pauli, C. & Reuser, K. (2009). *The Pythagoras study: Investigating Effects of Teaching and Learning in Swiss and German Mathematics Classroom The Power of Video Studies in Investigating Teaching and Learning* New York: Waxmann Publishing

Kinyanjui, P. (2004). *NEPAD initiatives in ICTs and Associated Capacity Building* Paper presented to the All Africa Ministers Conference on Open and Distance Learning, Cape Town. Feb, 2014. https://www.africaodi/conference/paper.htm

Kisser, M. & Wassmer, G. (1996). *Sample Size of a Pilot Study Using Upper Confidence Limit of the Standard Deviation* htts://www.uv.es>psicologicalPDF

Kisirkoi, F.(2015). Integration of ICT in Education in a Secondary School in Kenya: A case Study *Literacy Information and Computer Education Journal (LICEJ),6* (2)

Kiptalam, A. (2011). Challenges facing Computer Education in Kenyan Schools: ICT For the Developing World Available online http://ictworks.org/12challengesfacing

Klusmonn, U., Kunter, M., Trautwein, U. Ludtke, O. & Baumert, J. (2008). Teachers' Occupational well-being and Quality of Instruction: the Important Role of Self-regulatory patterns. *Journal of Educational Psychology100* (3)702

KNEC (2012) Examination Results Available on line at http://www-knec/cp/results

Kolb, D.A. (1991). *Learning Style inventory and Interpretation booklet* Boston, MA; McBer & Company

Kombo, K.D. & Tromp, A. L.D. (2006). *Proposal and thesis writing: An Introduction* Nairobi: Pauline Publications Africa

Kounenou, K., Roussos, P., Yotsidi, V. & Tountopoulous, M. (2015). Trainee Teachers Intention to incorporating ICT use into Teaching Practice in Relation to their Psychological Characteristics: The Case of Group Based Interventions. *Journal of Procedia-Social and Behavioural Sciences 190* 120-128

Kothari, C. R. (2004). *Research Methodology, Methds and Techniques* Wishwa Prakashan.

Krathwohle, D. R. (2003). *Educational and Social sciences, an Integrated Approach* Longman

Krejcie R. V. & Morgan D. W. (2000). *Determining Sample Size for Research Activities*, University of Minesota, Duluth www.opa.uprrp.edu

Kuo, Y. C., Walker A. E., Bellard B. R., Schrollar E.E. & Kou, T.K. (2014). *A Case Study of Integrating Interwise Interaction and Satisfaction in Synchronous online Learning Environments* available at www.researchgate.net/publication/277040

Laporan, P. (2004). *Chemistry Performance Report of SPM 2003* Malaysian Exam Board, Ministry of Education

Lawal, H. S. (2003). *"Teacher Education and Professional Growth of the 21st Century Nigerian Teacher"* The African Symposium 3-2

Law, N., Chow, A. & Allan, H. K. Y. (2005). Methological Approaches to Comparing Pedagogical Innovations using Technology *Journal of Education and Information Technologies 10* (2) 5-18

Loweless, A., Vootgt, G. D. and Bohlin, R. (2001). *"Something old, Something new. Is Pedagogy Affected by ICT?"* London: Routledge, 63-83

Legard, R., Keegan, J. & Ward, K. (2003). In-depth Interviews *Scientific Research Publishing Journal of Service science and Management 8* (3). https://www.scripts.

Lindfors, E. (2007). *ICT in Teaching: European Teachers Views.* University of Yurku https://www.researchgate.net/publication

Lochmiller, C.R. & Lester, N.J. (2017). *An Introduction to Educational Research:Connecting Methods to Practice.* Sage Publications: United Kingdom.

Lowther, D.L., Inan, F.A., Strahl, J.D. & Ross, S. M. (2008). Does Technology Integration work when Key Barriers are removed. *Journal of Educational Media International 45* 195-213.

Lu, Z. Hou, L. & Huang, X. (2013). A Research on a Student-Centered Teaching Model in an ICT Based English Audio Videos Speaking Class. *International Journal of Education and Development using ICT 6* 101-123

Luhamya, A. Bakkabulindi, F.E. K. & Muyindo, P. B. (2010). Integration of ICT in Teaching and Learning: A Review of Theories. https://researchgate.net.>2726.

Majumdar, S. (2005). *Regional Guidelines on Teacher Development for Pedagogy-Technology Integration* Bangkok: UNESCO-Asia and Pacific regional bureau.

Masagazi, J. Y. (2013). Getting Schools Ready for Integration of Pedagogical ICT: The Experience of secondary Schools in Uganda *International Journal of Academic Research in Business and Social Science* 3 (2) 2222-6990.

Maslowski, R. (2001). *School Culture and School Performance: An Explorative into the Organizational Culture of Secondary Schools and other Effects.* Twente University Press Enschede.

Merilyne, L. & Nobbert, P. (2006). *Learning to teach using ICT in secondary schools: A Companion to School Experience* http:/1/books.google.co.ke.

Microsoft (2017). *Unlock limitless learning for your Students.* https//Microsoft.com/en-us/education

Middleton, F. (2019). *Reliability and Validity: What's the Difference?* https://acc.13-1-2020

Mioduser, D., Nachmias, R., Lahav, O. & Oren, A. (1999). *Web-based Learning Environments (WBLE): Current Pedagogical and Technological*

States. Tel Aviv University, Science and Technology Education Center Research Report

Mishra,P. & Koehler M. (2011). *Technological Pedagogical Content Knowledge: A Framework for Teacher Knowledge* https://mkoehler.edu.msu.edu/tpack/category/core

Mishra,P., Haris, J. & Koehler, M. (2006). Technological Pedagogical Content Knowledge: A Framework for Teacher Knowledge and Learning Activity:Curriculum based Technology Integration Reframed. *Journal of Research on Technology in Education 4* (4)

MOE. (2012). *Task force on re-alignment of Education in Kenya:* Government Printers

MOE. (2006). *National Information and Communication Technology (ICT) Strategy for Education and Training.* Nairobi: Government Printers.

MOEST. (2005a). *Kenya Education Sector Support program 2005-2010; Delivering Quality Education and Training to all Kenyans.* Nairobi: Government Printers.

Morante, C. F., Lopez, B. C., Fernandes, J. C. & Miranda, T. M. (2014). *State of the Art of ICT Integration in European Schools Educational Systems: Equipment-Uses-Teacher Competencies* https://www.researchgate.net>272

Moroder, K. (2013). *Educational Technology Frameworks: Why I don't use TPACK or SMR with my Teachers.* http://www.edtechcoaching.org/2013/11/ed-tech.

Mugenda, O. M. &Mugenda, A.G. (2008). *Research Methods: Qualitative Approaches* Nairobi African Center for Technology Studies

Mugimu, B. C., Newby, S. L., Hite, M. J. & Hite, S. J. (2013). *Technology and Education: ICT in Ugandan Secondary school.* *https://www.researchgate.net>221*

Muhammad, A. U., Muhammad, M. A., Ahmed, S. A.A., Muhammed, M. R., AbdulKadar, M.M. & Nasrin, A. (2019). Impact of ICT on Students Academic Performance: Applying Association Rule Mining and Structured Equation Modeling *International Journal of Advanced Computer and Applications 10* (8)

Mukuna, T. (2013). Integration of ICT into Teacher Training and Development *Makerere Journal of Higher Education 5* (1)

Munyengabe, S., Yiyi, Z., Haiyan, H. & Hitimana, S. (2017). Primary Teachers' Perceptions on Implementation of One Laptop per Child (OLPC) Program in Primary Schools of Rwanda *EURASIA Journal of Mathematics, Science and Technology Education 13* (11) 7193-7204

Muriithi, M. J. (2017). *Factors Affecting Implementation of ICT Education in Public Primary Schools in Kajiado Sub-county, Kenya* Erepository.uonbi.ac.ke>Mu.PDF

Mwangi, M.I. & Khatete, D.(2017). Teacher Professional Development Needs for Pedagogical ICT Integration in Kenya: Lessons for Transformation. https://oapub.org>article>view.

Mwendwa, N. K. (2017). Perceptions of Teachers and Principals on ICT Integration in the Primary Schools curriculum in Kitui County, Kenya. *European Journal of Education Studies 3* (7)

Nacario, P. (2014). *Integrating UNESCO ICT-Based Instructional Materials in Chemistry Lessons* Central Bicol State University of Agriculture, Pili; Camarines.*Asia Pacific Journal of Multi-disciplinary Research 2* (4)

Nachmias, R., Mioduser, D., Cohen, A., Tubin, D. & Forkosh-Baruch, A. (2004). Factors Involved in the Implementation of Pedagogical Innovations using Technology. *Journal of Education and Information Technologies 9* (3) 291-308.

Narhe, P. (2005). *Code of Ethics: Plagiarism in Sinhggad Technical Education Society.* www.sinhgad.edu>PDF

Ndide, A.N. (2012). *Pedagogical Integration of ICT in Teaching and Learning in Educational Institutions in Uganda* Uganda-panAf-Report.pdf www.ernwaca.org

NEPAD. (2005). *A NEPAD e-Africa Commission Initiative for Africa* Nairobi: NEPAD e-schools, Kenya

Ngamo, T. S. (2006). Pedagogical Principles and Theories of ICT integration *AVU Teacher Education Authority Content workshop* Nairobi: Kenya.

Nilsen, T., Gustafsson, J.E. & Blomeke, S. (2016). Teacher Quality, Instructional Quality and Student Outcomes: Relationships across Countries, Cohorts and Time. *Journal of IEA Research for Education 2*

Nira, H. (1988). "Computer Based Drill and Practice in Arithmetic: Widening the gap Between High and Low-achieving students" *American Educational Research Journal 25* (3) 366-397

Nira, H. & Becker, H. J. (1994). "Problems and Potential Benefits." *International Journal of Educational Research 21* (1) 113-119.

Nkpa, N. (1997). *Educational Research for Modern Scholars Fourth Dimension Pulishers.*

Nolan, C. (2004). *What is Integral to Integration: Exploring Students Teachers Experiences and Understanding of ICT Integration* http://education.uregina.ca

Nouh, Z., Amer, N., Mostafa, N (2017). *Correlation between Students Perception of the learning Environment and their Academic performance* Available online a www.academia.edu/24590341/perceptions/htm

Nunan, D. (1992). *Research Methodists in Language Learning* Cambridge University Press

Odera, F. Y. (2002). *A Study of Computer Integrated Education in Secondary Schools in Nyanza Province, Kenya.* Unpublished PHD Thesis University of Pretoria: South Africa.

Ogini, O. I. & Popoola, A.A. (2013). *Effects of Mathematics Innovation and Technologyon Students Academic Performance in Open and Distance Learning (ODL).* Oasis.col.org/.../2013-Oginni%26Popoola Effects.mathematics.pdf

Okewa, J.J. (2011). ICT Adoption among Public Secondary Schools in Kisumu County, Kenya. E-respository.uobi.ac.ke>Abs

Olulube, N. (2005). "Appraising the Relationship between ICT usage and Integration and the Standard Teacher Education progress in Developing Countries". *International Journal of Education and Development 2* (3) 70-85 ijedict.dec.wwi.edu

Onwu, O.G. & Ngamo T. S. (2004). *Resources for Chemistry Educators* http://www.chem.com/chemed/digitexts Accessed on 13th July, 2013.

Onwu, O. G. & Ngamo, T. S. (2016). *ICT Integration in Education-Option Chemistry*. Available at https://oer.avu.org/handle/1234

Orodho, A. J. (2004). Elements of Education and Social Science: Research Methods. Nairobi: Masola Publishers

Osodo, J. (2010). *Potential for Incorporation of Computer Simulations in the Teaching of Geography in Secondary Schools* Unpublished PhD Thesis Maseno University

Oso, Y. W. & Onen, D. (2009). Writing Research Proposal and Report: A handbook for Beginning Researchers. Nairobi: Jomo Kenyatta Foundation

Ottevanger, W. (2001). *Teacher Support Materials as a Catalyst for Science Curriculum implementation in Namibia* Doctoral Dissertation, University of Twente, Enschede

Otieno, R.V. (2018). Influence of Teachers Status of Digital Literacy on Integration of Digital Technologies in Early tears of Education in Kenya *International Journal of Education and Research 6* (8). www.ijern.co>journal>A....pdf

Oyoo, M. A. (2018). Perception of Teachers on Influence of Motivational Strategies used by Career Teachers on Student Choice of Computer Studies in Secondary Schools in Kisumu County *International Journal of Education and Research 6* (8) Pandayechee, K. (2017). A snapshot survey of ICT Integration in South African Schools. *South African Computer Journal (SACJ) 21* (2) Grahams town

Pedro, F. (2012). *Trusting the Unknown: the Effects of Technology use in Education* In D. Soumitra and B. Bilbao-Osorio (eds). The Global Information Technology Report 2012: Living in a Hyper connected World. Geneva: World Economic Forum and INSEAD

Pelgrum, W.J. (2001). Obstacles to the Integration of ICT in Education: Results from World Wide Educational Assessment *Journal of Computer and Education 37* 163-178. Amsterdam: IEA

Pelgrum, W. J. & Law, N. (2009). *ICT in Education around the World: Trends, Problems and Prospects* International Institute for Educational Planninghttp://unesdoc.unesco.org/images/0013/001362/136281/e.pdf

Pleece, R. (2020). *College Park High School* https://www.instangram.com.pleece

Plomp, T., Pelgrum, J. & Law, N. (2007). SITES 2006 International Comparative Survayof Pedagogical Practices and ICT in Education *Journal of Communication and Information Technology 12* 83-92

Peters, R. (2010). *Pedagogies of Educational Transitions: European Antipodean* htpps: //books.google.co.ke>books

Rastogi, A. & Malhotra, S. (2013). *ICT skills and Attitudes as Determinants of ICT Pedagogy Integration* Availble at euacademic.org/upload article/22pdf

Razak, N. A. & Yahaya, M. A. (2002). The Classroom Teacher *Technology Journal of SEAMEO RESAM* vol 2

Roblyer, M.D., Edwards, J. & Harriluck, M.A. (2009). *Integrating Educational Technology into Teaching* (4th Ed.), Upper Saddle River, NJ: Prentice Hall.

ROK. (2013). *Vision 2030; 2nd Medium Term Plan 2013-2017: Pathway to Devolution, Socio- Economic development, Equity and National Unity* Nairobi:Government Printers   Available at http://www.infomation.go.ke/docs

ROK. (2013). *Education for Development* http://www.vvob.be/vvob

ROK, (2012). *National Research Report Kenya.* www.observatoire-tic-org

RoK, (2009). *Vision 2030; 1stMedium-Term Plan 2009-2013*

Ruales, P. & Andriano, T.C. (2011). Information and Communication Technology (ICT) Integration in Educational Institutions *Journal of Mindano State University Forum Vol.XXIV* (12).

Ryan, J. K. (1978). *The Belmonte Report: "Ethical Principles and Guidelines for the Protection of Human Subjects of Research"* DHEW Publication No (OS) 78-0012. US Government Printing Office Washington DC. 20402.

Sagra, A. & Gonzales, M. (2016). *The Role of ICT in Improving Teaching and Learning Process.* https://doi.org/10.1080/096877

Sanchez, C. J .J. & Aleman, E. C. (2011). Teachers Opinion Survay on the Use of ICT tolls to Suppport attendance Based Teaching. *Journal, Computers and Education 56* 911-915

Sanyal, B.C. (2011). *New Functions for Higher Education and ICT to Achieve Education for All Paper prepared for Expert Round Table on University and Technology* International Institute for Educational Planning UNESCO, Paris

Scrimshow, P. (2004). *Enabling Teachers Make Successful use of ICT* https://naivaproject,com>projects

Sekaram, U. (2003). Research Methods Business: A Skill Building Approach. Scientific Research Publishing *Open Access Library Journal 2*

Selinger, M. (2004). *Developing and Using Content in Technology enhanced Learning Environments: Globalization Trends in Science* University of BruneiGadong.

Siedel, T. & Shavels, R. J. (2007). Teaching Effectiveness Research in the past Decade: the Role of Theory and Research design in disentangling Meta analysis Results. *Journal of Educational Research 77* (4) 454-499.

Shahid, M. (2015). ICT Application Competency of Teachers: A Study to Understand the Level of ICT Application Skills and self Confidence among Science Teachers. http://worldconference.net/home *GSE-Journal of Education 3* 2015

Sharma, R. (2003). Barriers in Using Technology for Education in Developing Countries *Journal of Singapore Schools Computers and Education 41* (1) 49-63

Shiundu, J.S. & Omulando, S. J. (1992). *Curriculum: Theory and Practice in Kenya* Nairobi: Oxford University Press 1992. https://searchworks.stanford.edu.

Shuttleworth, M. (2012). Validity and Reliability Measurements Available at https://explorable.com/validity-and-reliability.

Steketee, C. (2006). Modeling ICT Integration in Teacher Education Courses using Distributed Cognition as a Framework [Electronic Version] *Australian Journal of Educational Technology 22* (1) 126-144

Stuart, C. (2015). *ICT Integration Models into Middle and High School Curriculum in the USA.* Adiyaman University, Adiyaman, Turkey

Sundipta, D. R. (2015). Application of ICTs in Teaching-Learning Process *International Research Journal of Multidisciplinary Studies 1* (7) 72-84Tearle, P. (2003). ICT Implementation: What makes the difference*British Journal of Educational Technology 34* (5) 567-583

Tezci, E. (2011). Turkish Primary School Teachers Perception of School Culture regarding ICT Integration. *Educational Technology Research Development 59* 429-443

Thayer, T. (2020). *Market Research insights* https://www.djsrearch.co.cu>item

Tong, K.P. & Trinidad, S.G. (2005). Conditions and Constraints of Sustainable Innovative Pedagogical Practices using Technology. *IEJFLL 93* (3) 1206-9620.

Tuju. R. (2004).National Information and Communication Policy Framework [A paper presented on 8th November 2004at Grand Regency Hotel, Nairobi]

UNESCO (2014) *ICT Enhanced Teacher Development Model* .UNESCO: Adis Ababa: UNESCO
11CBA2011http://www.eng.unesco.iicha.org/sites/default/files

UNESCO (2008) *ICT Competency Framework for Teachers* unescodoc.unesco.org/images/0021/002134/pc

UNESCO (2010) *Institute for Statistics Global Education Digest: Comparing Education Statistics Across the World.* Montreal : UIS

UNESCO (2004) *Integrating ICT into Education: Lessons Learned.* UNESCO: Bangok http://unesco.org/images/0012/001234/12348.2e.pdf

UNESCO (2004) *Teacher Education Guidelines: Using open and Distance Learning.* http://unesco.org/images/0012/001234/12348.2e.pdf

UNESCO. (2003). *Final Report: The workshop on the Development of Guideline on Teacher Training in Integration in Beijing China* UNESCO: Bangkok. http://unesco.org/images/0012/001234/12348.2e.pdf;

UNESCO (2002). School Networking's: Lessons Learned. *ICT Lessons Series, Vol. III.*

UNESCO. (2002b). *Information Communication and Technology in Education: A Curriculum for Schools and Program of Teachers Development* UNESCO: Bangkok. http://unesco.org/images/0012/001234/12348.2e.pdf.

Unwin, T. (2005). Towards a Framework for the Use of ICT in Teacher Training in Africa. *The Journal of Open, Distance and e-learning 20* (2) 113-129

Usun, S. (2009). Information and Communication Technologies (ICT) in Teacher Education (ITE) programs in the World and Turkey. (a comparative review). *Journal of Procedia and Behavioral Sciences 1* (2009) 331-334. https://unesco.org/images/0012/001234/12348.2e.pdf.

Vetta, L. & Getty, M (2018). Importance of Internet to Education Available on line at https://itstillworks.com/Importance-internet. Accessed on 11ᵗʰ October, 2018

Wabuyele, L. (2006). Computer use in Kenyan Secondary Schools: Implications for Teacher Professional Development In. C. Crawford *et al. Proceedings of Society for Information Technology and Teacher Education International Conference* (5) 2084-2090). Chesapeake, VA AACE

Wachiuri, R. N. (2015). Effects of Teacher Experience and Training on Implementation of ICT in Public Schools in Nyeri Central District, Kenya *IOSR Juornal of Humanities and Social Sciences 20* (3) 26-38

Wafula, J.M. & Wanjohi, N. (2007). *Kenya ICT Policy Draft Document* Nairobi:IDRC

Wallet, P. (2014). *ICT in Education in Asia* UNESCO institute of Statistics, Montreal, Quebec, Canada.

Wasike, A. J. (2018). Perceptions of Teachers, Learners and School Principals on the Integration of ICT in Teaching and Learning Agriculture in Bungoma County, Kenya. http://www.ir.library.egerton.ac.ke>per

Wastiau, P., Blamire, R., Kearney, C., Quiattre, V.E. & Monseur, C. (2017). *The Use of ICT in Education: a survey of Schools in Europe* https://orbi.uliege.be/bistreamweert, T.V. & Tatnall, A. (2005). *Information*

Communication Technology and Real Life Learning: New Education for New Knowledge Society. Springer, New York

Whitehead, A.L., Julious, S.A. & Campbell, M. J. (2020). *Estimating the Sample Size for a Pilot Randomized Trial to minim Overall Trial Sample size*

Sage Journals Life & biomedical Sciences https://journals.sagepub.com>abs

Willms, J.D. (2006). *Learning Divide:Ten Policy Questions about Performance and Equity of Schools and Schooling System* UIS Working Paper No. 5 Montreal: US

Wong, P. & Wettasinghe, M. (2003). *Managing IT Based Learning Environments.* In S. C. Tan & F.L.W (Eds)

Wong, Q. & Woo, H.L. (2007). Systematic Planning for ICT Integration *In-Topic Learning Journal in Educational Technology & Society, 10* (1) 148-156. http://www.ssta.sk.ca/research/technology/97-02.

Yamamoto, Y. & Yamaguchi, S. (2016). A Study on Teachers' self Efficacy for Promoting ICT Integrated Lesson in Primary Schools in Mongolia. *Journal of International Cooperation in Education 18*(2) 26-38

Yehudit, J. D., Rodrigues, S. & Sancha, S. (2014). How to Promote Chemistry learning through use of ICT-based Learning. *The Malaysian on-line Journal of science Education and Technology. 21* (2) 207-225.

Zang, H. R., Huang, Y.Z. and Zhang, J. (2012). *ICT and ODL in Education for Rural Development: Current Situation and Good Practices in China.* Beijing: UNESCO International Research and Training Center for Rural Education (INRULED) Beijing Normal University, R & D Center for Knowledge Engineering (BNU-KSE)

Zawescki-Richter, O., Musken, W., Krause, U., Alturki, U. and Aldraiweesh, A. (2015). Student Media Usage Patterns and Non-traditional Learning in Higher Education. *Journal of International Review of Research in Open and Distributed Learning 16* (2). https://doi.org/10.19173/irrodle.

Zimmer, E.J. & Carpi, A. (2003). *Vision Learning and Teaching Modules* http://www.visionlearning.com/lirary/teach.

APPENDICES

Appendix A: Questionnaire for School Principals

The purpose of this questionnaire is to gather information on the influence of ICT integration on student performance in chemistry in this school. Below are statements of availability, extent of use of ICT equipment and digital content, capacity building of teachers in the school. Please, where appropriate, use the keys provided to fill in the box or tick in the most appropriate category from the list of five alternatives.

Name of School:					
Number of streams:		Student population:			
Section A	**Availability of Computers, Telecommunication and Audiovisuals**				
KEY: number of equipment at ICT center: **1**-None, **2**-(1-3), **3**-(4-7), **4**-(8-9), **5**-(10 or more)					
State number of	**1**	**2**	**3**	**4**	**5**
Functional computers					
Students per computer					
communication system : mobile phones or e-mail					
DVD and CD players					
TV sets					
Lap tops or tablets for staff					
Video players					
OHP					
LCD projector					
Internet enabled computers					
Other (specify)					

Section B	The extent of use software in teaching and learning

KEY: 1: Not at all, **2:**To very little extent, **3:** To some extent**, 4:** To a great extent, **5:**To a very great extent

2. State the extent of use of:	1	2	3	4	5
Internet					
Content-free software					
Content-rich software					
E-libraries					
CD-ROM topical revision series					
DVDs for revision					
Websites					
CD-ROM encyclopedia					
Microsoft excel					
Microsoft office					
Learning systems					
Training software					
Audio shows					
Simulated video					
Animated video					
Subject videos					
Computer skills					
video shows					
Interactive learning systems					
Peer teaching					
Revision series					
Other (specify)					

3) State number of subjects with content in the following storage devices:

	0	1-3	4-6	7-9	10 and more
a)CDs					
b) Flash discs					
c) DVDs					
c) e-libraries					
d) CD-ROM discs					
e)other (specify)					

Section C: Teacher's capacity to use ICT	KEY: **1**-Not at all, **2**- BEd/Dip with less than 1 year certificate course in IT, **3**- BEd/Dip with up to 1 year certificate course in IT **4**-Diploma with IT, **5**- BEd.(Degree) with IT,

	1	2	3	4	5
4)Teachers ICT capacity What's your own ICT training level?	None	Certificate in IT	Diploma with IT	Higher National Diploma with IT	B.Ed. with IT

5)Number of teachers funded for ICT training:

	0	1-3	4-6	7-9	10 and more
a)Personal initiative					
b) BOG					
c) Ministry					
d) Donors					
e) Other (specify)					

Section D	**Perspectives of Influence of ICT use on performance of chemistry at KCSE and syllabus coverage**
State extent ICT integration has influenced syllabus coverage and has affected KCSE and chemistry **mean grade** in terms of **mean deviation** from time of inception	
KEY: 1:Not at all, **2**:To very little extent, **3**: To some extent, **4**: To a great extent, **5**:To	

a very great extent

	1	2	3	4	5

	1	2	3	4	5
6) How has ICT affected:	Negative	Not at all	+1	+2	+3
a)School syllabus coverage					
b) General KCSE school mean grade?					
c)KCSE chemistry mean grade					

*Thank you very much for your time, support and contribution.

Appendix B: Questionnaire for Chemistry Teachers

The effective integration of ICT into the school and classroom can both transform pedagogy and empower students. It is important that teachers are able to successfully weave technology into learning process. The following questions will help determine the level of your familiarity with ICT. Special attention is given to relevance of equipment, resource person's skills to use ICT in teaching and extent of usage of relevant equipment. Please fill in the information briefly or tick the most appropriate category from the list.

Name of school	
Section A	**Availability of computers, telecommunication, audiovisuals**

KEY: number of equipment at ICT center: **1**-None, **2**-(1-3), **3**-(4-7), **4**-(8-9), **5**-(10 or more)

1. State the number of:	1	2	3	4	5
Functional computers					
Students per computer					
Communication-system: mobile phones or e-mail					
DVD and CD players					
TV sets					
Lap tops or tablets for teaching staff					
Video players					
LCD projector					
Internet-enabled computers					
Other (specify)					

Section B	Frequency of use of software in teaching and learning				
KEY: 1: Not at all, **2:**Termly, **3:** Monthly, **4:**Weekly,**5:**Daily					
2) State the frequency of use of the following:	1	2	3	4	5
Internet					
e-libraries					
Video shows					
content-rich software in chemistry					
content-free software					
CD-ROM topical revision series					
Websites					
CD encyclopedia					
Microsoft excel					
Microsoft office					
Learning systems					
Training software					
Audio shows					
Simulated videos					
Animated videos					
Topical videos					
Computer skills					
Interactive learning software					
General revision series					
Other (specify)					

3) Approximate topics with digital content in the following storage devices:

	0	1-3	4-6	7-9	10 and more

a) CDs					
b) DVDs					
c) Flash discs					
d) E-libraries					
e) ROM discs					

Have you trained in ICT?

KEY: **1**-Not at all, **2**- BEd/Dip with less than 1 year certificate course in IT, **3**- BEd/Dip with up to 1 year certificate course in IT **4**-Diploma with IT, **5**- BEd.(Degree) with IT,

	1	2	3	4	5
ICT training					

Section C: frequency of Teacher's use of ICT skills in teaching chemistry

KEY: 1- Never, 2- Seldom, 3-Occassionally, 4- Frequently, 5-Always

	1	2	3	4	5
5) Can you browse by technology to navigate tutorials?					
6) Can you browse by task to find tutorials for particular teaching task?					

7) State the extent you are able to perform these tasks with ICT:
(a) In administration: Use the key:

1- Never, **2**- Seldom, **3**-Occassionally, **4**- Frequently, **5**-Always

	1	2	3	4	5
Compiling marks					
Creating work sheets					
Create a talking book					

(b) In lesson preparation: 1- Never, **2**- Seldom, **3**-Occassionally, **4**- Frequently, **5**-Always

	1	2	3	4	5

Animating diagrams					
Building web animation					
Downloading images					
Drawing a chart					
Drawing a table					
Inserting video into PowerPoint presentation					
Inserting pictures into an existing document					
c)In Teaching-learning:					
Conduct a lesson using interactive white boards					
Conduct class presentation					
Conduct an on-line class or discussion					
Create a quiz on PowerPoint using buttons and hyperlinks					
Structure lesson notes					
Teach students to write a report					

Section 4:Perspective of Influence of ICT on performance of chemistry and syllabus coverage

8) State extent ICT integration has influenced syllabus coverage and has affected KCSE and chemistry **mean grade** in terms of **mean deviation:**

KEY: 1: Not at all, **2:**To very little extent, **3:** To some extent, **4:** To a great extent, **5:**To a very great extent

	1	2	3	4	5
9.How has ICT affected:	**Negative**	**Not at all**	**+1**	**+2**	**+3**
a)Syllabus coverage					
b)General KCSE mean					

grade					
c) KCSE chemistry mean grade					

*Thank you very much for your time, support and contribution.

Appendix C: Questionnaire for Form 4 Chemistry Students

The purpose of this questionnaire is to gather information on your perception of the extent of implementation of ICT Integration Program offered in your school in teaching and learning chemistry as a subject. Read and understand each question before answering. Please tick in the box the appropriate choice or fill in your answer to the questions as briefly and as honestly as possible.

Name of your school	
Your class	

Section A : Availability of computers, telecommunication, audio visuals at ICT center

KEY: Number of equipment at ICT center: **1**-None, **2**-(1-3), **3**-(4-7), **4**-(8-9), **5**-(10 or more)

1. State the number of:	1	2	3	4	5
Functional Computers					
Student per computer					
Communication system: mobile phones or email					
DVD and CD players					
TV sets					
Laptops or tablets					
video players					
OHP					
LCD projector					
Internet enabled computers					
Other (specify)					

Section B Extent of use of software in teaching-learning learning chemistry					
KEY: 1-Not at all, **2**-To very little extent, **3**-To some extent, **4**- To a great extent, **5**-To a very great extent,					
2.State the extent of use of the following software	**1**	**2**	**3**	**4**	**5**
Internet					
Content-free software					
Content-rich software					
E-libraries					
CD-ROM topical revision					
DVDs for revision					
Websites					
CD-ROM encyclopedia					
Microsoft excel					
Microsoft office					
Learning systems					
Training software					
Audio shows					
Simulated video					
Animated video					
subject videos					
Computer skills					
Video shows					
Interactive learning systems					
Peer teaching					
Revision series					
Other (specify)					

3) To what extent is teaching done in these areas integrated with ICT digital content:

KEY: 1-Not at all, **2-**To very little extent, **3-**To some extent, **4-** To a great extent, **5-**To a very great extent,

	1	2	3	4	5
a)Spectrophotometer in titration at end points					
b)Duck builder in structure and bonding					
c)computerized molecular modeling and visualization					
d)ICT based chemistry lab					
e)Hot potatoes for quizzes					
f)PowerPoint presentations					
g)Live Internet sites					
h) Multimedia					
i)Movie clips in slides					
j)Other (specify)					
4)Suggest areas of difficulty that you as a learner found difficult to comprehend that have been improved by use of computers and ICT during teaching and learning					

*Thank you so much for your time, support and contribution.

Appendix D: Observation Schedule

Observation Schedule for use by the Researcher

This observation schedule will be used to do a physical check on the existing equipment, materials and resource persons in the study schools and will allow the researcher to determine the extent of ICT integration in the school, KCSE and chemistry mean grades.

Section A: School details:

Name of school..

Population of form 4 chemistry students in the school....................

Population of 4 chemistry teachers in the school.........................

1.   ICT Center staff Profile

Job description of ICT staff	Professional qualification	Experiences

Section 1:   Availability of computers, telecommunication and audio-visual

2.   Computer laboratory or ICT center

Available	
Not available	

3. Identify places where ICT is used in the school compound (List down)...
....
....

4. ICT store

Available	
Not available	

5. ICT technician

Available	
Not available	
Hired	

6. Places with computers (specify the number available)

Principal's office	
Secretary pool	
HODs office	
Career master office	
School ICT center	
School library	
Any other	

7. Other ICT equipment used in or outside the computer laboratory or ICT center

Equipment	Specific location
Telephone or mailing devices	
Video	
DVD and CD players	
TV sets	
Laptop or tablets	
Internet	
Any other	

8. List of equipment in the ICT center for teaching chemistry using ICT

Name of equipment	Quantity
Any other	

Section 2: extent of use of software programs in teaching and learning chemistry

9. List of storage devices with digital content for teaching chemistry using ICT

Nam Name of storage device	Quantity	Digdd Digital content present
CDs		
Flash discs		
DVD		
E-Libraries		
CD-ROM discs		
Other (specify)		

Section 3: Performance records

10.   School's chemistry results and KCSE examination results.

Year	Chemistry mean grade	No. of learners	School KCSE, mean grade	No. of learners
2011				
2012				
2013				
2014				
2015				
2016				
2017				

11. What is the distribution of chemistry students per grade in KCSE examination

	A	B	C	D	E
2011					
2012					
2013					
2014					
2015					
2016					
2017					

12. Observation of a lesson: Interactive dynamics between students and equipment in an ICT integration chemistry classroom session (An activity description lesson plan in the ICT integration). * See appendix E

Appendix E: Lesson Observation Checklist for ICT Integrated Lesson

School name...Time.........
Class......Class size..................Student computer experience (years)..............

1. Teacher Preparedness:

Teacher professional qualification	
Teaching experience (years)	
Computer Skill/ Training Level (Degree, Diploma, Certificate, Proficiency)	
No. of CEMASTEA In-set sessions attended	
Other ICT in-service courses attended e.g. ESP, IBM, Microsoft, School organized ICT In-Set or any other program (Please specify)	

2. Level of ICT integration of ICT resource in chemistry lesson

Key: ICT resources include Internet, e-library, video show, simulated video, animated video, Topical revision video, Content rich software, Topical revision series, Websites, Encyclopedia (tick appropriately type used)

Level of ICT integration in classroom:

S/No	Dimension		Component	None	Poor	Satisfactory	Good	Very good	Excellent
				0	1	2	3	4	5
1	**Preparation:** presence of technology integrated lessons in Schemes of work	1	Availability						
		2	Appropriateness						
		3	Adequacy						
		4	Timely						
		5	Relevancy & accuracy in concept						
		6	Realistic						
2	**Preparation:** presence of technology integrated lessons in Lesson plan	1	Availability						
		2	Appropriateness						
		3	Adequacy						
		4	Timely						
		5	Relevancy & accuracy in concept						
		6	Realistic						
3	**Extent of integration of ICT :** Name of Resource used (specify)……… ………….	1	Suitability & Adequacy						
		2	Variety & Originality						
		3	Relevancy						
		4	Utility & Effectiveness						
		5	Quality of Technological skill						
		6	Display						

4	**Level of use of ICT resource:** Method of delivery	1	Teacher demonstration						
		2	Class experiment						
		3	Expository approach						
		4	Inquiry based learning						
		5	Discussion						
		6	Project						
5	**Learner engagement with resource:** Learner activities	1	Attention drawn to resource						
		2	Observe teacher demonstration						
		3	Demonstrate concept with ICT						
		4	Experiment with ICT resource						
		5	Discuss observation from experiment						
		6	Make conclusions on observation						

Chemistry concept taught during integration……………………………..

Strength aspects……………………………………………………………

Weak aspects………………………………………………………………

Overall performance………………………………………………………

Suggestion for improvement……………………………………..Date……

The purpose of this questionnaire is to gather information on the (Economic Stimulus Program) ESP and other secondary schools ICT projects, its objectives and scope of operations. Kindly fill in your answers as briefly and as honestly as possible.

Name of Sub-county (Administrative District)................................

Personal details:

What are your:

1. Duties in school ICT project...

2. Qualification in ICT..

ICT School Project on computers, telecommunication and audio-visual:

3. How many computers were supplied to each school in the project?...
....
....

4. What is the ideal student: computer ratio that the ministry would wish to achieve?..

5. What is the current status of Student: computer ratio in ESP sponsored schools?..

6. How many CDs with digital content were donated to schools at the implementation of ICT integration in schools?...........................
....
....

7. In your opinion what should be the extent of use of ICT services by the following?
a) Administration...
b) Teachers..
c) ICT staff...
d) Learners...

Capacity Building of Teachers to use ICT skills in teaching

8. Which people at the school level do you liaise with?...

9. Which teachers do you work with in schools?..

10.   What kind of training do you give to the teachers?................................

11.   Whom do you use to train teachers?..

12.   How long do the teacher training sessions take?.................................

13.   Who finances the training of teachers on ICT skills?..........................

14.   Which policy guidelines does your project follow in training teachers?...

15.   In your opinion to what extent has ICT integration:
a) Enhanced syllabus coverage?..
b) Impacted on KCSE mean grade in ESP schools?................................
c) Impacted on KCSE chemistry mean grade?..

*Thank you so much for your time, cooperation and contribution

Appendix G: Letter of Introduction to School Principals

FROM: RUTH A. OTIANG'A
PHD STUDENT AT MASENO UNIVERSITY
TO: SCHOOL PRINCIPAL ....
Dear Sir/ Madam,
This is to inform you that I am the aforementioned, a researcher, researching on the
TOPIC: INFORMATION COMMUNICATION TECHNOLOGY PREPAREDNESS,
INTEGRATION IN EDUCATION AND STAKEHOLDERS PERSPECTIVES ON PERFORMANCE AMONG PUBLIC SECONDARY SCHOOLS IN KISUMU COUNTY, KENYA

Your school is one of the few, which have been sampled out for the study. It is for this reason that I request you to kindly complete the questionnaire that will be issued to you as accurately as you can. All the information you provide will be treated as confidential and used only for the purpose of which it is intended. None of the information will be published in a report in a manner which will enable any individual school, teacher or principal be identified.

I look forward for your cooperation in this exercise
Yours faithfully,

Ruth Atieno Otiang'a

FROM: RUTH A. OTIANG'A

PHD STUDENT AT MASENO UNIVERSITY

TO: CHEMISTRY TEACHERS OF..,

Dear Sir/ Madam,

This is to inform you that I am the aforementioned, a researcher, researching on the

TOPIC: INFORMATION   COMMUNICATION   TECHNOLOGY PREPAREDNESS,

INTEGRATION   IN   EDUCATION   AND   STAKEHOLDERS PERSPECTIVES ON PERFORMANCE AMONG PUBLIC SECONDARY SCHOOLS IN KISUMU COUNTY, KENYA

This research aims at determining the availability, extent of ICT integration, the capacity of chemistry teachers to integrate ICT and assess stake holders' perspectives on the influence of Information Communication Technology on chemistry performance in secondary schools in Kisumu County, Kenya.

Your school is one of the few, which have been sampled out for the study. It is for this reason that I request you to kindly complete the questionnaire that will be issued to you as accurately as you can. All the information you provide will be treated as confidential and used only for the purpose of which it is intended. None of the information will be published in a report in a manner which will enable any individual school, teacher or principal be identified.

I look forward for your cooperation in this exercise

Yours faithfully,

Ruth Atieno Otiang'a

Appendix I: Letter of Introduction to Chemistry Students

FROM: RUTH A. OTIANG'A

PHD STUDENT AT MASENO UNIVERSITY

TO: CHEMISTRY STUDENTS ....

Dear Master/ Miss,

This is to inform you that I am the aforementioned, a researcher, researching on the

TOPIC: INFORMATION COMMUNICATION TECHNOLOGY PREPAREDNESS,

INTEGRATION IN EDUCATION AND STAKEHOLDERS PERSPECTIVES ON PERFORMANCE AMONG PUBLIC SECONDARY SCHOOLS IN KISUMU COUNTY, KENYA

Your school is one of the few, which have been sampled out for the study. It is for this reason that I request you to kindly complete the questionnaire that will be issued to you as accurately as you can. All the information you provide will be treated as confidential and used only for the purpose of which it is intended. None of the information will be published in a report in a manner which will enable any individual school, teacher or principal be identified.

I look forward for your cooperation in this exercise

Yours faithfully,

Ruth Atieno Otiang'a

Appendix J: Letter of Introduction to SCQASO

FROM: RUTH A. OTIANG'A
PHD STUDENT AT MASENO UNIVERSITY
TO: SCQASO…………………………………………………..…….sub-county
Dear Sir/ Madam,
This is to inform you that I am the aforementioned, a researcher, researching on the
TOPIC: INFORMATION COMMUNICATION TECHNOLOGY PREPAREDNESS,
INTEGRATION IN EDUCATION AND STAKEHOLDERS PERSPECTIVES ON PERFORMANCE AMONG PUBLIC SECONDARY SCHOOLS IN KISUMU COUNTY, KENYA

Your school is one of the few, which have been sampled out for the study. It is for this reason that I request you to kindly complete the questionnaire that will be issued to you as accurately as you can. All the information you provide will be treated as confidential and used only for the purpose of which it is intended. None of the information will be published in a report in a manner which will enable any individual school, teacher, principal or sub-county quality assurance and standards officer be identified.

I look forward for your cooperation in this exercise

Yours faithfully,

Ruth Atieno Otiang'a

Appendix K: Information for Consent to Participate by Respondents

You are hereby informed of a research titled: ICT Preparedness, Integration in Education and Stakeholders Perspectives on Chemistry Performance among Public Secondary Schools in Kisumu County, Kenya. That is to be conducted at your institution. The purpose of this research is to investigate ICT preparedness, integration in education and stakeholders perspectives on student performance in chemistry. You have been identified as a participant in this important social research program. The information given by you will be used to determine the relevance of the structure of ICT integration and the information may be used by curriculum developers in recommending appropriate methodology for ICT integration in secondary schools. The following will help you understand the exercise:

- You will be expected to respond to questions in the questionnaires administered to you. Respond as honestly as possible and to the best of your knowledge.
- There are no known risks that will come from your participation, on the other hand you will benefit by being a source of information to this important exercise.
- The identity of your institution and yours as an individual will remain anonymous and any information given by you will be confidential.
- You are free to ask any questions on issues you don't understand and we will respect your decision in case you decline to participate in the research.
- However, once you have consented to participate in this exercise, your full cooperation is expected.
- In any case if you latter decide to withdraw from the research, your decision will still be respected; consequentially your institution will miss out on the contribution to this important exercise.
- Under age learners (aged below 18 years) otherwise referred to as minors will be included in the research. However their consent to participate will be sought from their teachers who are expected to have their best interests at heart.

This research is conducted and supervised by the following:
1. Ruth Atieno Otiang'a,   2. Dr. Mildred A. Ayere,
    P. O. Box 3923-40100,   P.O. Box 19182-40-123,
    Kisumu, Kenya.   Kisumu, Kenya
    Tel. 0724-846494.   Tel. 0724-374433
3. Dr. Joseph A. Rabari,
    P.O. Box 333-40105
    Maseno, Kenya.
    Tel. 0721-524786

This research is sponsored and partially funded by:
Maseno University,
P.O. Private Bag,
Maseno, Kenya.
Tel. (056) 351222/351008/351011

This section should be completed by the respondent.

I have read and understood the contents of this information and I voluntarily consent to participate in this research as frankly and as honestly as I can

Name_________________________________signature__________________Date

Appendix L: Research Authorization

MASENO UNIVERSITY

SCHOOL OF GRADUATE STUDIES

Office of the Dean

Our Ref: PG/PHD/00024/2012

Private Bag, MASENO, KENYA
Tel:(057)351 22/351008/351011
FAX: 254-057-351153/351221
Email: sgs@maseno.ac.ke

Date: 12th October , 2017

TO WHOM IT MAY CONCERN

RE: PROPOSAL APPROVAL FOR RUTH ATIENO OTIANG'A — PG/PHD/00007/2010

The above named is registered in the Doctor of Philosophy in Pedagogy programme in the School of Education, Maseno University. This is to confirm that her research proposal titled "Level of Teacher Preparedness, Use and Influence of Information Communication Technology Integration on Student Performance in Chemistry among Public Secondary Schools in Kisumu County, Kenya" has been approved for conduct of research subject to obtaining all other permissions/clearances that may be required beforehand.

Prof. J.O. Agure
DEAN, SCHOOL OF GRADUATE STUDIES

183

Appendix M: Approval by MUERC

MASENO UNIVERSITY ETHICS REVIEW COMMITTEE

Tel: +254 057 351 622   Ext 3050
Fax: +254 057 351 221

Private Bag – 40105, Maseno, Kenya
Email: muerc-secretariate@maseno.ac.ke

FROM: Secretary - MUERC

DATE: 2nd May, 2018

TO: Ruth Atieno Otlang'a
      PG/PHD/0007/2010
      Department of Educational Communication,
      Technology and Curriculum Studies
      School of Education, Maseno University
      P. O. Box, Private Bag, Maseno, Kenya

REF: MSU/DRPI/MUERC/00488/17

RE: Proposal Reference Number MSU/DRPI/MUERC/00488/17 Extent of Teacher Influence, Preparedness Use and Influence of Information Communication Technology Integration on Student Performance in Chemistry among Public Secondary Schools in Kisumu County, Kenya.

This is to inform you that the Maseno University Ethics Review Committee (MUERC) determined that the ethics issues raised at the initial review were adequately addressed in the revised proposal. Consequently, the study is granted approval for implementation effective this 2nd day of May, 2018 for a period of one (1) year.

Please note that authorization to conduct this study will automatically expire on 1st May, 2019. If you plan to continue with the study beyond this date, please submit an application for continuation approval to the MUERC Secretariat by 15th April, 2019.

Approval for continuation of the study will be subject to successful submission of an annual progress report that is to reach the MUERC Secretariat by 15th April, 2019.

Please note that any unanticipated problems resulting from the conduct of this study must be reported to MUERC. You are required to submit any proposed changes to this study to MUERC for review and approval prior to initiation. Please advice MUERC when the study is completed or discontinued.

Thank you.

Dr. Bonuke Anyona.
Secretary,
Maseno University Ethics Review Committee.

Cc: Chairman,
Maseno University Ethics Review Committee.

MASENO UNIVERSITY IS ISO 9001:2008 CERTIFIED

"""

Appendix N: Research Permit

MINISTRY OF EDUCATION, SCIENCE AND TECHNOLOGY
State Department of Education

Telegrams:"schooling",Kisumu
Telephone: Kisumu 057 - 2024599
Email: Ckisumu@education.go.ke

When replying please quote

COUNTY DIRECTOR OF EDUCATION
KISUMU COUNTY
PROVINCIAL HEADQUARTERS NYANZA
3RD FLOOR
P.O BOX 575 – 40100
KISUMU

CDE/KSM/GA/19/3A/(129)

5th February 2016

TO WHOM IT MAY CONCERN

RE: RESEARCH AUTHORIZATION
OTIANG'A RUTH ATIENO – REG. NO.PG/PHD/0007/2010

The above named is a student at Maseno University.

This is to certify that she has been granted authority to carry out research on *"Using Information Communication Technology to Improve Performance in Chemistry in Kisumu County, Kenya "* for a period ending August, 2016.

Any assistance accorded to her to accomplish the assignment will be highly appreciated.

SILVESTER MULAMBE
COUNTY DIRECTOR OF EDUCATION
KISUMU COUNTY

Appendix O: Kisumu County Secondary Schools Enrollment

MINISTRY OF EDUCATION
State Department of Early Learning and Basic Education
KISUMU COUNTY

S/No.	SUB COUNTY	CATEGORY	NO. OF SCHOOLS	ENROLLMENT Boys	ENROLLMENT Girls	TOTAL
				2019 SECONDARY SCHOOLS ENROLLMENT		
1	KISUMU CENTRAL	Regular	12	5083	4268	9351
		Special	1	193	207	400
		Private	8	306	471	777
		Total	21	5582	4946	10528
2	KISUMU EAST	Regular	15	3197	3172	6369
		Special	-	-	-	-
		Private	4	174	274	448
		Total	19	3371	3446	6817
3	KISUMU WEST	Regular	36	7362	6830	14192
		Special	-	-	-	-
		Private	2	147	39	186
		Total	38	7509	6869	14378
4	MUHORONI	Regular	33	5739	6637	12376
		Special	1	57	53	110
		Private	4	313	530	843
		Total	38	6109	7220	13329
5	NYAKACH	Regular	52	8501	11367	19868
		Special	1	69	63	132
		Private	1	123	146	269
		Total	54	8693	11576	20269
6	NYANDO	Regular	42	7109	7238	14347
		Special	-	-	-	-
		Private	3	173	112	285
		Total	45	7282	7350	14632
7	SEME	Regular	35	5529	6461	11990
		Special	-	-	-	-
		Private	-	-	-	-
		Total	36	5529	6461	11990
	COUNTY TOTALS	Regular	225	42520	45973	88493
		Special	3	319	323	642
		Private	22	1236	1572	2808
		Total	250	44075	47868	91943

Appendix P: Krejcie and Morgan Table and Formula

Table 3.1

Table for Determining Sample Size of a Known Population

N	S	N	S	N	S	N	S	N	S
10	10	100	80	280	162	800	260	2800	338
15	14	110	86	290	165	850	265	3000	341
20	19	120	92	300	169	900	269	3500	346
25	24	130	97	320	175	950	274	4000	351
30	28	140	103	340	181	1000	278	4500	354
35	32	150	108	360	186	1100	285	5000	357
40	36	160	113	380	191	1200	291	6000	361
45	40	170	118	400	196	1300	297	7000	364
50	44	180	123	420	201	1400	302	8000	367
55	48	190	127	440	205	1500	306	9000	368
60	52	200	132	460	210	1600	310	10000	370
65	56	210	136	480	214	1700	313	15000	375
70	59	220	140	500	217	1800	317	20000	377
75	63	230	144	550	226	1900	320	30000	379
80	66	240	148	600	234	2000	322	40000	380
85	70	250	152	650	242	2200	327	50000	381
90	73	260	155	700	248	2400	331	75000	382
95	76	270	159	750	254	2600	335	1000000	384

Note: N is Population Size; S is Sample Size   *Source: Krejcie & Morgan, 1970*

Formula for determining sample size

$$s = X^2 NP(1 - P) \div d^2 (N - 1) + X^2 P(1 - P)$$

s = required sample size.

X^2 = the table value of chi-square for 1 degree of freedom at the desired confidence level (3.841).

N = the population size.

P = the population proportion (assumed to be .50 since this would provide the maximum sample size).

d = the degree of accuracy expressed as a proportion (.05).

Source: Krejcie & Morgan, 1970

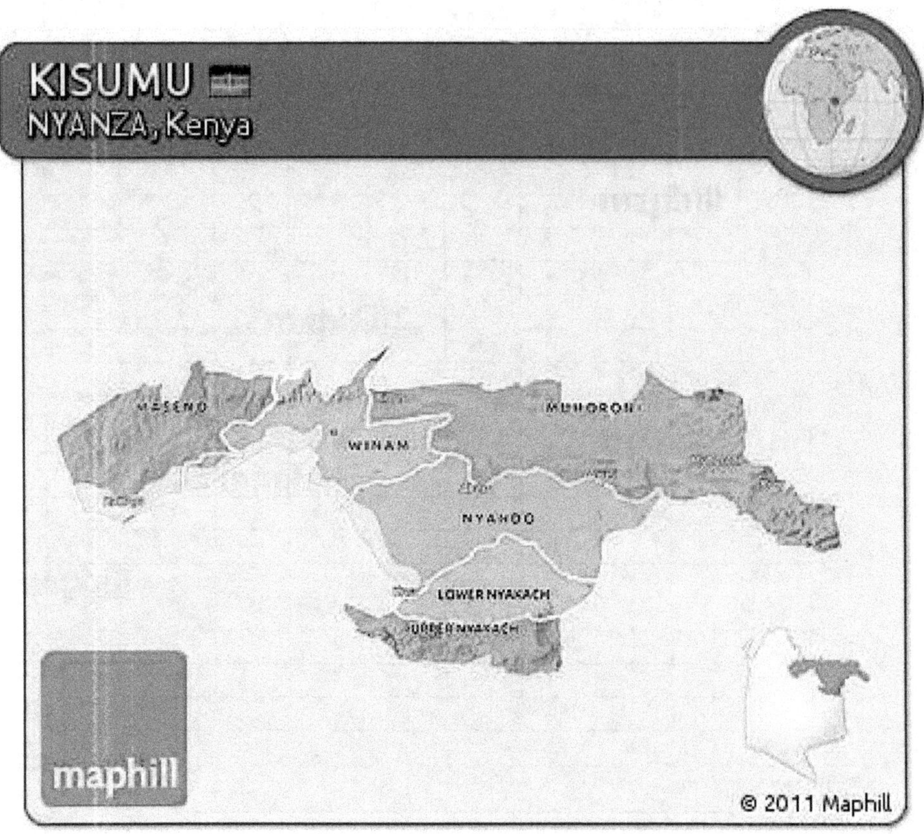

KISUMU
NYANZA, Kenya
MASENO
WINAM
MUHORONI
NYANDO
LOWER NYAKACH
UPPER NYAKACH
maphill
© 2011 Maphill